Air-conditioning America

Johns Hopkins Studies in the History of Technology
Merritt Roe Smith, Series Editor

Air-conditioning America

Engineers and the

Controlled Environment, 1900–1960

Gail Cooper

The Johns Hopkins University Press

Baltimore and London

© 1998 The Johns Hopkins University Press
All rights reserved. Published 1998
Printed in the United States of America on
acid-free recycled paper

9 8 7 6 5 4 3 2 1

The Johns Hopkins University Press
2715 North Charles Street
Baltimore, Maryland 21218-4319
The Johns Hopkins Press Ltd., London

Library of Congress Cataloging-in-Publication
Data will be found at the end of this book.

A catalog record for this book is available
from the British Library.

Design by Christine Taylor
Composition by Wilsted & Taylor Publishing Services

ISBN 0-8018-5716-3

To my parents
Dr. C. D. and Paula Cooper

Contents

Acknowledgments

I wish to thank those institutions that kindly lent me access to their archives and collections: Cornell University, United Technologies, American Society of Heating, Refrigerating, and Air-conditioning Engineers, Smithsonian Institution, Harvard University, Theater Historical Society, National Association of Home Builders, and the University of California at Santa Barbara. In particular I would like to thank Anne Milbrooke at United Technologies for her help with the Carrier records.

My research in those collections would not have been possible without financial support, first from the Smithsonian Institution, where I was a fellow, and then from Lehigh University, where I joined the history faculty.

This book began as a graduate dissertation at the University of California at Santa Barbara. I wish to thank my committee there—Carroll W. Pursell, Patricia Cline Cohen, Henry D. Smith, and Sarah Fenstermaker Berk—for their help and encouragement. I owe an enormous debt to Carroll Pursell, who has acted consistently as a friend and mentor.

In addition to funding, the Smithsonian's National Museum of American History also provided an energizing place to work. It was there that I first met colleagues who have provided critical support and helpful criticism: Robert Vogel, Robert Post, Steven Lubar, Robert Freidel, Susan Smulyan, Colleen Dunlavy, and Carolyn Cooper. Others generously read and critiqued the manuscript: my colleague at Auburn University, Lindy Biggs, patiently read *x*

repeated drafts of my first article, as did Philip Scranton, while Hans Dienel, Joe Corn, and Bernard Nagengast read the entire manuscript. Many strengths of the book come from their suggestions, while any failings that remain are entirely my own.

Although teachers, colleagues, and friends have provided timely help toward the completion of the dissertation, the first article, and the first book, none can match the contribution of my family, for their support has been continuous through every stage of this manuscript. I am grateful for the loving support of my parents, C. D. and Paula Cooper, and for the early encouragement from my siblings, Cathy, Susan, and Chris. Most crucial to the successful completion of this book, however, has been the intelligence, wit, and endless patience of my husband, Bruce Sinclair, for which I am endlessly grateful.

Introduction

This book centers on the debate, sometimes acrimonious, over open windows: the ones that let in cool breezes and the smell of spring flowers, summer evenings, and damp autumn leaves; the same ones that let in city noise, smokestack pollution, street dirt, and hot, humid air. For years both the pleasant and the irritating aspects of opening windows were inevitably linked, for the weather has always been something admittedly beyond human control. That is why we still laugh at Mark Twain's joke that "everyone talks about the weather, but no one does anything about it." Yet with the appearance of air conditioning, the technical community began to take seriously the idea of creating a man-made indoor climate—the mechanical reproduction indoors of the best aspects of the weather outside. At last, engineers argued, human beings would be in control. We could—and some argued that we should—close our windows forever.

Only with the appearance of air conditioning did engineers and architects believe that a totally artificial indoor environment, independent of the natural climate, was a possibility. Until the early decades of the twentieth century, artificial climates were the stuff of science fiction novels; indeed, Rosalyn Williams has pointed out that a man-made environment was central to many fictional accounts of an advanced society that reached beyond its natural and primitive beginnings.[1] Although winter heating was one of the first steps in the mechanization of buildings, it was never accompanied by the option or necessity of permanently sealing off the inside from the outside, as air-conditioning engineers suggested. So until the development of air conditioning, buildings were semi-permeable barriers, in which we could not only look out the window to see the natural landscape but also open that window to let in the sensual delights of fresh air.

Such an appreciation of the open window was expressed by playwright Horton Foote in a recent interview. Foote has made a career out of recounting his seven decades of life in the small town of Wharton, Texas, a community that built its prosperity on cotton, corn, and sugar cane. He remembers that "when the cotton mills were running full-time and they had a cotton-seed mill, we would have this wonderful odor permeating the house. I find myself thinking, 'What was that really like and why did it vanish?' I don't hear the train whistles like I used to. That haunting lonely sound. We would be sitting on the porch and hear the different churches and the flats where the blacks had their restaurants. I don't hear that anymore. Every place is air-conditioned. We don't keep the windows open anymore."[2] Foote is not alone in lamenting a changed community life resulting from air-conditioned houses. Experts tell us that many of our memories are linked to smells, so perhaps it is inevitable that we should feel nostalgic about a recollected past of open windows.

However, within many factory buildings successful processing depended on a stable and consistent atmospheric environment, achieved by a close control over the levels of temperature and humidity. Any open window defeated that goal, and air-conditioning contracts often stipulated that windows remain closed. That impetus toward substituting mechanical ventilation for open windows culminated in the 1950s and 1960s, when many new buildings featured sealed windows. The debate over the virtues of natural versus man-made climate focused on the attributes and desirability of buildings that were dependent upon mechanical services to provide healthy and comfortable atmospheric conditions.

This tension between the use of open windows or mechanical systems to provide ventilation emphasizes that the history of air conditioning is the history of air, and not, as modern consumers might assume, of cooling. For the first three decades of the twentieth century, air-conditioning installations were built sometimes with refrigeration machinery and sometimes without. Refrigeration was not considered an essential feature of the technology. These were formative decades in which a small coterie of engineering experts maintained technical and commercial dominance over the new technology. Out of that close-knit community emerged a specific design for air-conditioning systems and a vision of air conditioning's proper role that placed primary importance on mechanical precision, engineering control, and independence from natural climate—elements captured in the term *man-made weather.*

Early air-conditioning systems required that buildings—and, consequently, people's activities—be organized around technical requirements. The question became, then, not simply whether to air-condition buildings, but what form the technology would take and who would determine that configuration. The public discussion was widely based and included workers, factory owners, public health officials, laboratory researchers, school reformers, electrical manufacturers, public utilities, and homeowners. Although most people in these groups were not qualified to critique the design of air-conditioning systems in terms of the mechanics of condensers and compressors, they showed a keen interest in such elements of atmospheric control as recirculation, air volume, temperature, and humidity. Thus, the debate ranged from abstract issues, such as the proper relationship between mechanical civilization and nature, to practical concerns, such as schoolroom odors.

The legacy of this struggle is two distinct traditions in the deployment of air conditioning. One is the choice of design professionals, engineers and architects, who favor a controlled and rational system, a building that is so integrated with its mechanical services that it becomes a machine itself and is controlled by technical authority. A second is the choice of some users, who want an interior that is more comfortable but not necessarily ideal and who favor a technology that is above all flexible and responsive to the consumer's needs. The first is represented by the powerful central air-conditioning systems found in modern hotels and office buildings, which most nearly achieve the ideal of man-made indoor weather; the second is represented by the affordable and portable window air conditioner with its small blast of cold air.

This split in air-conditioning technology has resulted in a giant public policy headache, for each approach has structural defects that came to light in the 1960s and 1970s. There are many people who detest the centralized control of the sealed, air-conditioned building that deprives them of choice over their immediate environment. At the same time, the machine-in-the-window approach to interior comfort can truthfully be described as engineering lunacy. It fails to integrate the building and the machine, thus contributing to the general problems of the global energy crisis. Many Americans personally experience the schizophrenic character of modern air conditioning by working in one kind of air-conditioned space and living in another. Because both corporate

4

and personal choices have such far-reaching impact, it is important to consider whether technological design can be wholly entrusted to either the technical elite or the market forces of our consumer culture.

This study is founded on the belief that material culture shapes everyday life and that technological choices have important consequences. It focuses on the development of air conditioning as a series of contested choices and not as an inevitable progressive development. Convinced by Langdon Winner that technology has politics, I have looked at the ways in which the technical community built assumptions about standards, use, and values into air-conditioning systems, and how other interest groups opposed the engineers' vision.[3] While I have viewed engineers and engineering firms as the primary shapers of new technology, I have tried always to place them in the commercial context in which they worked. Perhaps the most useful insight of this study is the importance of the commercial setting in allowing the original vision of technical designers to reach consumers. Custom production gave engineers greater control over the design and use of the technology than did mass production. Custom design in industrial air-conditioning systems and in large-scale comfort installations more thoroughly integrated technical expertise and the cultural values embedded in engineering standards with the product. I believe this will prove characteristic of custom production in general and not only of the air-conditioning industry in particular.

The interest in how technology reflects the values of society is not new. A vibrant tradition in the social history of technology has recently acquired a more theoretical approach with the growth of scholarly interest in the social construction of technology. The social-constructionists ambitiously include a vast network of actors that shape the development of technology in society. These "actors" are defined variously as people, institutions, economic factors, social values, and technical resources, and the sprawling, comprehensive character of their analysis often precludes discussion of the disproportionate power of some actors.

This study is more specifically focused on social groups and the power relationships among them. Specifically, it examines the relationships among engineers, consumers, and corporations. I believe that engineers have a privileged role in the construction of technology, but that the practice of engineering nearly always occurs within an economic context and a corporate culture. Although engineers and business firms may remain the most influential shapers of technology, consumers have considerable power as well. Consumer preferences have long been considered an integral part of Adam Smith's invisible hand that guides the marketplace; they are the most important part of "demand" in a supply-and-demand economy. But I have tried to transcend a view of consumer behavior as the practice of economic democracy in which buyers vote with their dollars. Consumers can indeed buy or boycott, but they can also sabotage, regulate, cajole, and demand—or simply create a pattern of usage that resembles not at all the vision of designers. This book focuses on the ways in which engineers, consumers, and corporations use a variety of tactics in the seesawing power relations surrounding the development of air conditioning, including rhetoric, models, government regulation, guarantees, industry standards, experimental science, and quantitative standards.

Each chapter of this study examines an important market for air conditioning in the period from 1902 to 1960, where the interests of these three groups came into conflict: factories, schools, theaters, and homes. Technically, these sites cover only two types of air conditioning: process air conditioning, represented by factories, and comfort air conditioning, represented by schools, theaters, and homes. From another perspective, these four markets also fall into two categories: custom-designed systems for factories, schools, and theaters, and mass-produced appliances for the home. Yet they represent four distinct groups of consumers with different budgets, standards, and viewpoints, so they are treated separately in this study.

Chapter 1 lays out the vision of three pioneering engineers for a new type of comprehensive ventilation system called air conditioning that incorporated the ability to control humidity, and which thus promised to create artificial indoor climate, or man-made weather. Chapter 2 examines the adoption of process air conditioning among seasonal manufacturers and the resulting three-way struggle among engineers, workers, and managers for control of the factory floor. Chapter 3 sketches the conflict over mandatory mechanical ventilation in public schools and the resulting development of an experimentally defined comfort zone and of quantitative standards for the ideal climate. Chapter 4 explores the legacy of the ventilation debate in the fledgling industry of comfort air conditioning. In particular, it analyzes the acceptance by theater architects, film exhibitors, and motion-picture audiences of conceptions about man-made weather, the ideal climate, and the comfort zone. Chapter 5 depicts the upheaval caused by the development of the window air conditioner and traces the introduction of mass-production methods in pursuit of the residential market during the Depression. The declining power of the engineer-designer under mass-production methods signaled the weakening of the model of air conditioning as artificial climate. Chapter 6 returns to the residential marketplace to suggest that the widespread adoption of comfort air conditioning after World War II was built less on the mass-produced air conditioner than on the mass-produced tract house. In the postwar years air conditioning became a means for underwriting a new style of construction and architectural design in both homes and office buildings. Chapter 7 discusses the acceptance of air conditioning by consumers.

Last, let me say a word about sources. The papers of Willis Carrier and the companies he helped establish are an important source for this study, yet I have tried to draw a portrait of the industry rather than of one individual or a handful of firms. The relevance of those records to the larger story stems partly from the exceptional nature of the man and his companies, and partly from their representative character. As a pioneer and leader within the air-conditioning industry, Carrier and the companies he founded often participated in technical developments or professional movements that were of great importance to the evolving industry. Certainly, Carrier and his companies were outspoken about their vision of the new technology. They alternately represented and shaped the views of the industry. Yet as distinguished as their role in the industry was, their records reflect the difficulties that all custom-design firms experienced during the reshaping of the industry to include mass production. In that way Carrier Engineering Corporation was not exceptional, but instead illustrated the problems of all the early engineering firms that were suddenly faced with competition from the manufacturing giants.

CHAPTER ONE

It's Not the
Heat, It's the Humidity

Most modern consumers equate air conditioning with cooling. Air conditioning is for the summer months when heat forces people to seek refuge in the cool indoors. In fact, nothing so thoroughly shows the distance between air conditioning's origins and its current state as that redefinition of the term that describes it. For the early engineers who pioneered its development, air conditioning meant control of humidity levels. Beginning in the first decade of the twentieth century, air-conditioning engineers argued that humidity regulation was an important part of modern mechanical ventilation systems, whether those installations were for industrial processing or human comfort.

Their claims had a practical basis. Many manufacturing processes were sensitive to changing humidity levels, and, in fact, the control of humidity led to better quality, fewer defects, and the opportunity for year-round production. Human beings are sensitive to humidity as well. The increased moisture in winter and the drier conditions in summer made possible by air conditioning helped people feel more comfortable. But the new technology emerged as more than a simple response to consumer demands. Many in the industry promoted and sold air conditioning as artificial indoor climate. Once engineers found the means to regulate humidity, they claimed that alongside the other technical marvels of the modern era, they had finally achieved mastery over the weather as well. With the addition of humidity regulation to temperature control and air circulation—the essential ingredients of air conditioning as they defined it— engineers could reproduce the best of natural climate indoors and achieve a complete separation of the indoors from the outdoors.

The history of air conditioning's early success, then, involves both the pull of existing consumer markets and the push of a technocratic vision of perfect man-made weather. Both the market and the ideal focused on the importance of humidity control. It was the central premise of man-made weather, and when designers and consumers alike embraced the possibilities of artificial indoor climate, they faced the necessity of closing off the indoors from the outside weather in order to maintain that control.

EARLY COOLING EXPERIMENTS

All technologies have a prehistory, a partially realized stage of evolution, and air conditioning is no exception. One of the earliest visionaries to articulate the benefits of atmospheric cooling was Florida physician John Gorrie, who focused on the importance of temperature rather than humidity. As early as 1842 Gorrie proposed "to counteract the evils of high temperature, and improve the condition of our cities, [by] the rarefaction and distribution of atmospheric air previously deprived of large portions of latent caloric by mechanical condensation."[1] In other words, he envisioned atmospheric cooling for health and comfort, through the deployment of mechanical refrigeration. Initially, Gorrie's aims were linked to his medical interests. He hoped to neutralize the deleterious effects of tropical climate: malarial disease and the enervating effects of heat and humidity. His vision included not only mechanically cooled houses but also a central refrigeration plant for an entire city. Grand as these ideas were, it is unclear whether he ever actually built a cooling system. By 1844 he had constructed a refrigeration machine, which he used for ice production.

In 1854 Gorrie's vision of the potential benefits of refrigeration had expanded beyond comfort cooling to include manufacturing. He believed that "the cheap and abundant production of artificial cold would enable many industrial arts to be carried on advantageously in warm climates, or continuously in temperate regions." Gorrie's argument that refrigeration would be of value in moderate climates as well as in hot and humid regions was later echoed by many proponents of air conditioning.

The earliest experiments in the first stages of this technology incorporated cooling. By 1889 central-station refrigeration was commercially available in several cities in the United States and abroad. Subscribers to these services enjoyed refrigeration piped into their buildings by central-station systems similar to those that provided electricity, gas, and water. Refrigeration companies laid pipes in the streets, usually in a restricted business district, to carry refrigerant to nearby subscribing customers, where it was used primarily for cold storage.[2] Central-station refrigeration systems were built in New York, Boston, Baltimore, Los Angeles, St. Louis, Denver, and Kansas City.

With the ready availability of mechanical refrigeration, at least one customer experimented with comfort cooling. In St. Louis, Missouri, the St. Louis Automatic Refrigerating Company supplied refrigeration to the Ice Palace, a restaurant and beerhall on Market Street, between Third and Fourth Streets. In 1891 the enterprising restaurant devised a comfort-cooling system by attaching three banks of expansion coils to the wall halfway between the floor and ceiling. The arrangement was reported to work very well. "No difficulty is found in maintaining any desired temperature in the room, no matter how crowded it may be or how warm the air outside," reported the trade magazine *Ice and Refrigeration*.[3] The room temperature often registered between 72 and 75 degrees when outside conditions hit 85 or 90. To capitalize on this pleasant effect, the Ice Palace decorated its walls with paintings of the Kane Polar Expedition and other snowy scenes. The expansion pipes that carried the refrigerant from the central station spelled out the proprietor's name in the front window where they became heavily covered with frost, a further enticement to customers.

More typical of the early air-cooling experiments was the use of ice and fans. In 1907 ventilating engineer John J. Harris was inspired to place ice cakes in the air-supply ducts of a school auditorium during the graduation ceremonies.[4] Many ephemeral installations like his were probably used and not recorded. Ice cooling offered great flexibility, since it could be used on special occasions without the standing expense of a costly refrigeration plant. Yet ice cooling was bulky, awkward, messy, and never really economical for long-term use. None of these attempts at room cooling incorporated the systematic control of humidity levels, which came to distinguish air conditioning.

Three engineers played particularly important roles in building sophisticated mechanical systems independent of the outside climate, and in promoting the possibilities of controlled indoor environments: Alfred Wolff, Stuart Cramer, and Willis Carrier. Before 1918 all three had designed and built mechanical ventilation systems that incorporated humidity regulation, thereby pioneering the development of man-made weather.

ALFRED WOLFF

In 1893 Dr. R. Ogden Doremus lauded the wonders of modern chemistry and physics in an article in the *North American Review*. Doremus, a faculty member of the College of the City of New York, was particularly enthusiastic about the use of mechanical refrigeration to cool banks, businesses, homes, and hospitals. "If they can cool dead hogs in Chicago, why not live 'bulls and bears' in the New York Stock Exchange?" he asked rhetorically.[5] Doremus was not the first to advocate indoor cooling, but he may have been the most prophetic. Indeed, in 1902, when the New York Stock Exchange built a new building between Broad and New Streets with a wing extending to Wall Street, the trading room was equipped with a cooling system as well as a heating system. But this first installation did not mark the beginning of a trend, and cooling remained uncommon for a number of years.

The Stock Exchange cooling system was as unusual as its designer, the innovative engineer Alfred R. Wolff. Wolff was born in Hoboken, New Jersey, on 16 March 1859. He studied engineering at Stevens Institute of Technology, graduating in 1876 at the age of seventeen. He then entered the office of engineer Charles E. Emery as an unpaid apprentice. He was thus the product of both the newer and the older forms of engineering education. Wolff spent two years in Emery's office, then two more as a second assistant engineer in the U. S. Revenue Marine Service. Not until 1880 did he set up his own office in the Astor House in New York City at the age of twenty-one. He described himself as a steam engineer.[6]

Wolff made his reputation in 1888, when he designed the heating and ventilating plant of the New York Freundschaft Club. It was an unqualified success, and it launched Wolff on his lifelong interest in heating and ventilating systems. One of his friends conceded that in 1888 this seemed "a single and comparatively limited line of engineering work," but Wolff was at the forefront of a new engineering specialty. Indeed, enough practitioners were engaged in heating and ventilating engineering by 1904 that they formed a separate professional society, the American Society of Heating and Ventilating Engineers (ASH&VE). Wolff soon had more engagements than he could complete singlehanded, and he spent his later years supervising a large office of assistants.[7] By the time Wolff was tapped for the New York Stock Exchange job, he was a leading practitioner of the new art in New York City. Still, the use of refrigeration for comfort cooling seemed more theoretical than practical to those designing the new Exchange. Wolff convinced the building committee of its potential, but the nervous chairman, Mr. Potts, warned him that if the installation proved unsuccessful, Wolff had better buy a one-way ticket to Europe.[8]

Predictably, Wolff's first experiments with comfort cooling had involved ice and fans. In 1889 he designed the ventilation system for Carnegie Hall, equipping the air-supply ducts of the concert hall with racks for ice for use during hot weather. His associate doubted that the hall—or the system—was ever used in the summer.[9] But Wolff increasingly became aware of the importance of humidity control in building a successful comfort installation.

The relationship of humidity levels to indoor comfort had long been recognized. Expert and layman alike knew that cold, damp weather chilled one to the bone, while hot, humid days made people feel sticky and uncomfortable. Cold, clammy air wicked away body warmth, while a hot, steamy atmosphere prevented the cooling evaporation of sweat from the skin. This basic recognition of humidity's importance to comfort was followed by an understanding of its relationship to temperature. Temperature and humidity are necessarily connected, and any change of the indoor temperature affects the humidity level as well.

The room temperature establishes a natural limit to the amount of moisture the air can carry before it becomes saturated. Cold air has a small capacity for moisture, while hot air has a large capacity. The quantity of moisture in the air, measured in grains, establishes the absolute humidity level. Humidity levels can also be indicated as relative humidity—that is, the percentage of the air's maximum capacity. Since maximum capacity changes with temperature, the relative humidity is specific to temperature.

The most frequent complaints about indoor comfort are dry air in the winter and humid air in the summer. Cold air, with its small capacity for moisture, has a low absolute humidity; thus, when it is brought indoors and heated, the relative humidity plummets as the temperature rises. The resulting hot air is dry, and it dehydrates both people and furnishings. In her famous housekeeping manual, published in 1873, Catherine Beecher recommended placing a container of water on the stove to combat this problem, with the expectation that evaporation would raise the relative humidity to more comfortable levels.

The drying effect of woodstoves was soon surpassed by the dry heat of steam and hot-water radiators. Occupants complained that these systems produced "canned air," an imprecise term that suggests the smell of hot, dry, metallic air. One ventilating engineer likened the indoor atmosphere during the winter heating season to the desert environment. Hot, dry air pulled the moisture from human mucous membranes, skin, and hair. That is why so many people found the air of modern buildings enervating and unpleasant.

Wolff's sophisticated handling of the control of humidity for comfort distinguished several of his installations. Although his contemporaries understood the importance of humidity and had a sketchy understanding of its behavior, few had developed techniques to control it. As he contemplated the task of cooling the New York Stock Exchange, Wolff incorporated features from a series of installations he had completed between 1893 and 1902. During these years, he had developed a set of characteristic practices for control of temperature, humidity, cleanliness, and distribution of air—the four essential functions of modern air conditioning.

Wolff used an uncommon but effective method of calculating the heating requirements for each installation, which gave him accurate control of temperature. He adopted the "heat-unit system" for calculating the size of heating installations, a method that was mandated for government buildings in Germany and widely employed there but little used in the United States.[10] Under this system engineers established the size of radiator surfaces for direct heating based on the proportion of window surface and the estimated heat loss through various types of wall construction. This improved upon a common American method of estimating the square footage of radiator surface based on the cubic content of the room, regardless of the construction of the building. This American method, one engineer complained, "repeatedly resulted in huge failures."[11] In contrast, the heat-unit system reflected a more sophisticated understanding of the relationship between the mechanical services of a building and its architectural features. That lesson of interdependence was one that air-conditioning engineers learned over and over again.

In 1893 Wolff began using a split-distribution system, which involved a plenum, or mixing chamber, of tempered air for ventilation and direct radiation to provide heating. In this way he could regulate each function separately. In addition, a combined plenum and exhaust system more effectively circulated the air through the building. The next year he adopted cheesecloth filters to ensure the cleanliness of the incoming air supply. Last, Wolff turned to winter humidification. In 1893, when he had designed the heating and ventilating system in the Cornelius Vanderbilt residence, he had used a compressed-air thermostatic control produced by Johnson Control Company to regulate the heat. He encouraged the company to adapt the control to humidity regulation, and in 1902 he installed a winter heating system in the Andrew Carnegie residence that not only provided humidification to combat the dry heat of winter, but also boasted automatic control of humidity levels using the Johnson equipment. Only an extremely successful practitioner had the opportunity to use homes of the very wealthy, such as Vanderbilt and Carnegie, as an experimental laboratory.

Wolff's interest in the effect of humidity control on comfort led him to conduct experiments with the ventilation system he designed for the new Cornell Medical College building in New York in 1899. Although he had already designed a successful system for winter heating and humidification, the Cornell installation was his first venture into cooling and dehumidification. This project provided important data that Wolff used for the New York Stock Exchange design.

The Cornell Medical College building, on First Avenue between Twenty-eighth and Twenty-ninth Streets, had a unique cooling system on the fifth floor. It was not strictly a comfort-cooling system. Since it served the postgraduate dissecting room, we may presume that its cool temperatures helped preserve the cadavers during classroom study. The installation used a large ammonia-absorption refrigeration machine built by the Carbondale Machine Company. The rating of any refrigeration machine was expressed by the number of tons of ice necessary to produce an equivalent cooling effect in a twenty-four-hour period. Thus, the "50-ton" Carbondale machine supplied for the Cornell Medical College installation provided a cooling capacity equal to melting 50 tons of ice in a day. In a classic fan-coil arrangement, the refrigeration machine cooled a brine solution that was carried by pipes—in a classic fan-coil arrangement—to the fifth floor, where air blown over the pipes produced a supply of cold air. Fresh air was taken from 10 feet above the roofline and drawn by fans through a cheesecloth filter providing 600 square feet of surface. The air could either be taken directly into the postgraduate dissecting room without cooling or be forced through a mass of brine coils in the attic before it entered the room. Stale air was removed through four exhaust registers, a pair of which could be closed for recirculation to increase the cooling effect. These exhaust registers and the damper controlling the bypass around the cooling coils were operated automatically by a thermostat.

Although the system was designed for the special needs of the dissecting room, the school administrators could not resist the temptation to use it for comfort cooling as well. Any time that crowds assembled, their collective body heat could easily make the indoors far hotter than the outside weather. So, following the building's opening, graduation exercises were held in the main dissecting room on the fifth floor, and the door to the postgraduate dissecting room was left open to provide cooling for the 1,500 people who attended. Records were kept of the performance of the machinery upon that occasion, and one can imagine Wolff carefully noting what would be necessary to build a comfort-cooling installation. At 12:30 on the afternoon of graduation, the outside temperature was 88 degrees, the temperature in the postgraduate room was 62 degrees, and that of the main dissecting room was 90 degrees, slightly above the outside temperature. At 3:00 P.M., the outside temperature was 87 degrees, the temperature in the postgraduate dissecting room was 66 degrees and that of the main dissecting room was 88 degrees, just one degree higher than the outside temperature. That was the best differential between inside and outside temperature achieved in the main room that day. Calculations of the work done by the cooling coils in the attic space over the ceiling yielded an estimate that "a total refrigerating capacity of about 205,000 heat units [BTU] per hour are required, that is about 17 tons [of refrigeration] under conditions which are perhaps not the severest, for a room the cubical contents of which are about 17,000 cubic feet."[12] Of special interest are Wolff's calculations that at conditions of 71 percent humidity, more energy is required to cool the moisture in the air than is required to cool the air itself. Here we can see that the engineer was calculating how much capacity was required to bring down the humidity as well as the temperature.

Wolff drew upon all this experience for the New York Stock Exchange installation. The Exchange system provides a revealing example of Wolff's engineering style. The outside air was drawn in through cheesecloth filters. The heating system was an indirect system, controlled by Johnson automatic temperature regulation. The amount of radiating surface was "determined on the lines usually followed by Mr. Wolff"—that is, the German heat-unit method.[13] It was supplied with a plenum and exhaust system, with the exhaust exceeding the supply to provide a positive movement of air to the outside. The winter humidity was supplied by steampipe coils submerged in a water bath to set up evaporation, with an automatic water-feed valve for keeping the water level above the coils.

While all features of the building's heating and ventilation system represented Wolff's standard notions of good engineering practice, the comfort-cooling installation far outstripped in complexity his 1889 ice-cooling system for Carnegie Hall. The air-cooling system "is the most important feature of the mechanical installation," *Engineering Record* wrote, "as it marks the introduction of a provision for comfort during hot weather that is the leading example of its kind in existence both from its magnitude and the exacting conditions of its service."[14] The cooling system used three ammonia absorption machines from the Carbondale Machinery Company, providing a total of 450 tons of cooling capacity. It was a fan-coil system; that is, a calcium chloride brine was cooled by the refrigerant and circulated through a cooling coil over which the air was circulated by fans. In this respect it resembled the Cornell Medical College installation.

What marked this installation as different was the recognition of humidity as a prime cause of summer discomfort, the establishment of 55 percent humidity as a standard for which to strive, and an informed calculation of the capacity necessary to achieve that level. Wolff aimed at maintaining 75 degrees Fahrenheit and 55 percent relative humidity. To size his equipment to produce that effect, he took into consideration the average temperature and humidity levels from the records of the local U.S. Weather Bureau and the heat gain through walls and windows, from the occupants of the room, and from the electric lamps. He calculated that he would need 119 tons of cooling capacity for cooling the air and 180 tons for extracting the moisture from the trading room. Although Wolff's system did not include a method for measuring the humidity and adjusting the machinery accordingly, it can perhaps be called the first air-conditioning installation because of its self-conscious attempts to control all four key factors: temperature, humidity, cleanliness, and distribution of air.

An article in a New York newspaper suggests that Wolff placed a "similar system" in the Hanover National Bank that used 12,000 feet of brine coils for cooling.[15] However, the New York Stock Exchange installation did not have many imitators. The system that Wolff installed was a response to the construction of ever larger buildings whose increasingly complex mechanical plants made them in some ways machines in themselves. Wolff was aware of the extent to which the new mechanical ventilation installation was inextricably bound up with a whole complex of technical innovations. In 1894 he noted that the general introduction of electric lighting and the generating plants necessary to run them provided the perfect complement to heating and ventilating plants. "It is an interesting fact," he observed, "that the quantity of steam required for heating and ventilating large buildings . . . equals closely the amount used independently for their electric lighting."[16] The heating and ventilating systems could efficiently be run off the exhaust steam produced from the generation of electric lights. This made the large building an independent unit in terms of power consumption. "It is this fact," Wolff explained, "which makes it difficult for either city or district heating or electric lighting companies to supply steam or electricity to large buildings. They cannot compete."[17]

In this Wolff was mistaken. The view of the modern building as an independent machine was superseded by an architectural perspective that classed such technological features as mechanical services.

Wolff's career as a consulting heating and ventilating engineer blossomed at a time when architects were beginning to acknowledge their reliance on increasingly complex technology. Lighting, telephones, elevators, heating, and ventilating all contributed to making the "skyscraper" livable. Large buildings required the expertise of the engineer in addition to that of the architect. Engineer Reginald Pelham Bolton complained that such recognition was not happening fast enough. "The architectural profession as a body have failed to realize their responsibility in regard to this matter of the employment of proper engineering skill upon the design and proportion of the power plant of these large buildings," he told an audience of mechanical engineers.[18] As Wolff himself explained,

> It is no longer unusual for our large hotels, office buildings and auditoriums, to contain a boiler plant of 1,000 horse-power, with a corresponding electric lighting, heating, cooling and refrigerating system, nor for the equipment of these plants to require an outlay of hundreds of thousands of dollars. And with these large machinery interests at stake, it becomes important that the skill of the trained engineer be called into play to ensure, in co-operation with the architect, the most desirable and best attainable results, with a judicious expenditure of money.[19]

Wolff maintained working relationships with several architectural firms, most notably with George B. Post and Sons. He collaborated with Post not only on the New York Stock Exchange, but also on buildings for the College of the City of New York (1905), the Mutual Benefit Life Insurance Company (1906), and the Cleveland Trust Company (1907), as well as the Cornelius Vanderbilt residence (1893). He developed strong credentials as an engineer who could deliver a successful system, and he did so by combining a willingness to adopt new practices with an insistence on conservative standards of quality that encompassed few risks. A contemporary noted that he "made it a matter of principle, not to undertake any work unless it could be planned with a view to permanency, efficiency, and economy; and he always impressed upon his clients the fact that the first cost of such equipment should receive less consideration than the cost of operation and maintenance."[20] Wolff's well-known insistence on quality was an expensive practice, and as a consequence he acquired an impressive list of wealthy clients. He installed only three residential systems: one for Cornelius Vanderbilt, one for Andrew Carnegie, and one for J. J. Astor.

It is no accident, then, that *Engineering Record* noticed the convergence in the New York Stock Exchange of big money, beautiful architecture, and innovative engineering. "The new building of the New York Stock Exchange," it wrote, "besides being conspicuous as the central home of American finance and one of the architectural adornments of the city, contains an engineering equipment of exceptional value."[21] Although the cost of this early system prevented its widespread adoption, it was clearly influential. One engineering journal was undoubtedly thinking of the work of Alfred Wolff when it wrote:

> Plans of heating systems in modern mercantile buildings, such as banks, and in more costly residences, are sure to delight the eyes of the designing engineer, as well as all right-minded contractors. In these classes of buildings are to be found, perhaps, the most perfect installations and the most strictly up-to-date practices, with those little touches of luxury that cause the systems to do many works of superogation which, while not absolutely needful, are, nevertheless, appreciated and have their effect in establishing ideals for the profession to live up to.[22]

This, then, was the beginning of air conditioning, before a name had even been found to describe it. There were few customers, however, with the wealth sufficient to support such systems, and for years air conditioning remained a luxury. Although their number was increasing, expensive and sophisticated modern commercial buildings provided a small venue for the development of air conditioning. Yet Wolff's work was important in establishing the idea that humidity control—both winter humidification and summer dehumidification—was an integral part of the modern heating and ventilating system.

STUART CRAMER

At roughly the same time that Wolff pioneered air conditioning for comfort among the New York elite, Stuart Cramer developed air conditioning as an aid to processing in the growing textile industry of the American South. Industry provided a larger market for air conditioning than luxury residences and expensive commercial buildings, and it was this larger market that sustained the early technical innovations and their developers. In industry as well as in commercial buildings, humidity control was combined with heating and ventilat-

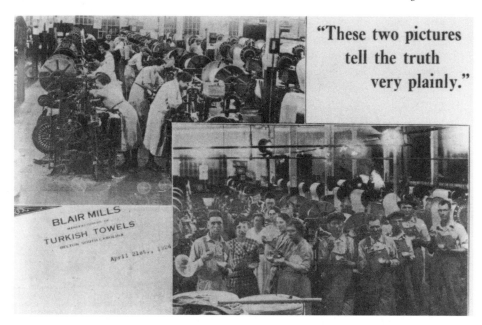

"These two pictures
tell the truth
very plainly."

BLAIR MILLS
TURKISH TOWELS

April 21st, 1924

Textile mill workers. In the top photograph, textile mill workers are busily tying up broken ends caused by excessive dryness. In the bottom photograph, they are enjoying a dish of ice cream after the installation of humidifiers. One of the first markets for air-conditioning systems was in industrial plants to control humidity for processing. (Southern Textile Bulletin 26 [19 June 1924]: cover)

ing equipment to produce a new system that had a greater range of functions along with much more complex technology.

Process air conditioning—the term used for these industrial applications—began with the effort to add moisture to dry factory environments. Humidification was especially important in factories that processed hygroscopic materials—that is, materials that absorb moisture from the air, which changes their shape and working properties. For example, manufacturers took advantage of the natural tendency of cotton and wool to absorb moisture, for damp fibers were more elastic and tough, easier to card, spin, and weave. Atmospheric humidification had the additional advantage of keeping static electricity levels low. Older methods of atmospheric humidification included dampening the factory floor or adding steam to the atmosphere. These methods were supplemented in the 1870s by commercial humidification systems, which used compressed air to atomize water over the looms.

No one gave more thought to the problems of humidification in textile mills than Cramer. A textile mill engineer from Charlotte, North Carolina, he coined the term *air conditioning* in a paper read before the National Cotton Manufacturers Association in May 1906, using it to describe his recent development of a new kind of textile mill humidification. Cramer recalled how "when entering this field, several years ago, I was puzzled to find a word that would embrace this whole subject. In casting about, I finally hit upon the compound word, 'Air Conditioning,' which seems to have been a happy enough choice to have been generally adopted."[23] The term was suggested by the use of *yarn conditioning*, which referred to the practice of exposing the textile fibers to moist atmospheric conditions in storage rooms before processing. Rather than condition the materials, Cramer proposed conditioning the air to which they were exposed as they were worked.

Cramer's term *air conditioning* emphasized atmospheric humidification over direct methods of moistening. He was groping for a new word because his equipment differed from existing systems of humidification, both direct and atmospheric. Rather than striving for the simple goal of adding moisture to the air, he hoped to maintain a predetermined relative humidity. Cramer aimed to substitute a precise numerical percentage of relative humidity for rough quantitative values, such as "more." Textile workers did not always agree on the optimum humidity level for a specific task. However, the ability to control the factory at a given percentage of humidity allowed the manufacturers to avoid large swings in humidity levels that often characterized older systems. Cramer's air-conditioning system maximized control and consistency.

Cramer was born in 1867 in Thomasville, North Carolina, and graduated from the U.S. Naval Academy at Annapolis, Maryland, in 1888. He resigned from the navy that year and attended Columbia University School of Mines from 1888 to 1889. For the next four years he worked at the U.S. Assay Office at Charlotte. In 1893 Cramer began his career in textile engineering, joining the office of D. A. Thompkins Company in Charlotte as engineer and manager. When one of Thompkins's clients, the Whitin Machinery Works, became discontented with the company's handling of their account, Cramer left Thompkins to become Whitin's southern sales agent. He eventually represented several textile machinery manufacturers of which Whitin was one.[24] He also established several independent businesses, including the Cramer Furniture Company and the Cramer Air Conditioning Company.

The Cramer Air Conditioning Company grew out of Stuart Cramer's long involvement with the problems of textile manufacturing. From his position as the agent of several textile machinery companies, he designed or equipped nearly one-third of the South's cotton mills from 1895 to 1918.[25] This deep engagement with the problems of textile manufacturing led him to cast about for ways to improve factory humidification. Changes in mechanized weaving only increased the importance of atmospheric humidification. Increased machinery speeds led to hotter rooms and more highly stiffened yarns. In 1911 the *Southern Textile Bulletin* noted that "in these days of automatic high speed looms, it is necessary to size heavily" and that "unless the room is heavily moistened, it is impossible to soften up a hard sized warp" in the short time that it was exposed to the atmosphere on the frame.[26]

In 1904 Cramer applied for his first patent in air conditioning, a control instrument for regulating humidifiers automatically. It was the first of many inventions; at the time of his death in 1940, he held sixty patents that reflected the broad range of his interests.[27]

Cramer's 1904 control instrument, a hygrometer, was based upon the principle of a constant wet-bulb depression. Ventilating engineers ordinarily used two kinds of thermometers. One, a dry-bulb thermometer, was the kind commonly found in homes to measure heat. The second, a wet-bulb thermometer, measured the temperature of the air when cooled to saturation. It thus reflected not only heat but also the relative humidity of the atmosphere. A fabric wick covered the bulb of the wet-bulb thermometer, and was kept constantly wet by submerging one end in a small reservoir of water. The evaporation of the moisture from the wick cooled the thermometer so that the wet-bulb temperature was always below the dry-bulb temperature unless the air was completely saturated.

Cramer's control consisted of a wet-bulb and a dry-bulb thermometer wired electrically in such a manner as to maintain a constant difference between the wet-bulb and dry-bulb temperatures. That constant difference, or *wet-bulb depression,* maintained the relative humidity of the atmosphere at a nearly constant percentage. A 6-degree wet-bulb depression at a temperature of 70 degrees produces 69 percent relative humidity, while the same depression at 90 degrees produces 73 percent relative humidity. Thus, the variation in humidity was only 4 percent over a range of 20 degrees. "In practice," Cramer wrote, "it is found that within ranges of temperature ordinarily encountered in a textile factory a regular and uniform depression of the 'wet bulb' (as it is termed) corresponds with a practically-uniform relative humidity."[28] In addition, Cramer patented an electrically controlled compressed-air valve that provided automatic regulation of a humidifier when linked to the hygrometer.[29] Between June 1905 and April 1906 he filed patents on four humidifiers as well.[30]

Textile mill. Bell-shaped air conditioners designed and installed by Stuart Cramer hang from the ceiling in the weaving room of a textile mill. They resemble the humidifiers that they replaced. (Cramer Air Conditioning Company, *Equipment* [n.d., n.p.])

Not only did Cramer's system provide automatic control of humidity, increasingly he merged ventilation and humidification into one system. The humidifiers that preceded Cramer's air conditioner were canister-shaped devices that hung from the ceiling every few feet and aspirated a fine mist of water that mixed with the air of the workshop more or less effectively, leaving any excess to settle on the equipment. Such a constant wetting of the machinery promoted rust. Cramer's first air conditioners were similar in appearance and were mounted on the ceiling in great numbers. However, in these devices the air of the factory was drawn inside the housing, conditioned, and then released to the atmosphere of the shop, thus eliminating excess moisture that fell on the machinery.

In this fashion Cramer made the switch from providing moisture to providing conditioned air. This new approach to an old problem led him to manufacture a new air conditioner that could be mounted on the wall to draw in fresh air and condition it before circulating it within the factory. This model, unlike previous humidifiers, was designed to perform the functions of both ventilation and humidification. It was placed against an outside wall or window and equipped with a fresh-air intake. Air was drawn both from the outdoors and from the factory and pulled through a fine water spray and then through a cloth filter before being discharged. Cramer claimed his air conditioner both cleaned and humidified the air.

Although the device was clearly aimed at solving the problem of maintaining an adequate humidity by adding moisture, the patent application claimed that it would also act as a condensing or cooling tower to lower the humidity in some cases. On those days when the humidity rose to a "very high percentage," Cramer expected a portion of the moisture to precipitate onto the cold surfaces of the cloth filter. Whether it would actually dehumidify in any effective fashion is unclear, but within a limited range, his April 1906 air conditioner cleaned, humidified, and distributed the entire air supply for the inside of a factory, doing so automatically when coupled with his humidity control instruments. This combination of functions was what Cramer termed "air conditioning," and together with temperature control, it remained the basic definition of the technology.

Cramer argued that the air conditioner offered manufacturers greater control over atmospheric conditions. The combination of ventilation and humidification in one apparatus permitted the replacement of window ventilation with a mechanical system. Manufacturers could now keep factory windows closed. Windows produced adequate ventilation only at the risk of creating drafty conditions. In his 1906 patent application Cramer asserted that "it is a well-known fact that ventilating textile factory buildings by opening windows or doors is not only injurious from a manufacturing standpoint interfering with the proper running of work, but also positively disarranges and disturbs the normal conditions of the fibers of the material which are required for the most favorable conditions to manufacturing."[31] By 1906 he arrived at a closed-window strategy for manufacturers of hygroscopic materials. His combination of humidity control and mechanical ventilation systems was the beginning of the separation and isolation of the factory interior from outside conditions, for the purpose of maintaining a precisely controlled indoor climate. With the adoption of this system, Cramer worked toward an ideal that he expressed as "the most favorable conditions to manufacturing," rather than the older goal of the improvement of existing conditions.

Air conditioning thus developed from two different sources: a concern for human comfort and an aid to industrial production. While comfort air conditioning preceded process air conditioning, it had a limited market. Both, however, involved the effort to control humidity, in addition to concerns of the temperature, cleanliness, and distribution of air. The emphasis on dehumidification for comfort led Wolff to adopt a fan-coil system with full refrigeration equipment, while Cramer's concern with humidification led to the use of a spray chamber. Both engineers went beyond the current engineering practice of their day in attempting to design a comprehensive system of environmental control using both existing and innovative equipment. Attempts to match technical equipment to these increasingly elaborate notions of environmental control absorbed the energies of the industry for years.

WILLIS CARRIER

The pioneer who did the most to promote the acceptance of a precisely controlled indoor environment and the standardization of equipment around the spray chamber was Willis Carrier. Carrier was a Cornell University graduate with a degree in mechanical engineering, whose first job in 1901 with the Buffalo Forge Company of Buffalo, New York, set his professional identity as a "centrifugal fan man," a specialist in fans and ventilation.[32] The company first tried to make a salesman out of him, but the man who drew this assignment, J. Irvine Lyle, sent Carrier back to the home office in 1902 with the recommendation that he be given research and development work.[33] This early assessment of Carrier's talents proved extremely accurate. Later associates told numerous stories of an absent-minded theorist—of the time, for instance, that he volunteered to make dinner reservations in the hotel dining room but had forgotten his mission by the time he arrived there and instead sat down to eat by himself, leaving his party of friends waiting endlessly in the hotel lounge. Carrier's absentmindedness was the flip side of his talent for analytical thinking; thus, in an industry that rewarded engineering skill with business success, his idiosyncrasies were markers of the personal strengths he brought to the development of this new technology. Carrier has often been called the "father of air conditioning"; his preeminence in the industry is best understood in terms of his gift for combining theory and practice, the tenacity of his commitment to this emerging field of engineering, and his business success.[34]

One of Carrier's important contributions was to give air conditioning an institutional home in the series of firms that bore his name.[35] The first was the Carrier Air Conditioning Company of America, a wholly owned subsidiary of the Buffalo Forge Company, formed in 1908 for the design and sale of air-conditioning systems. Buffalo Forge was owned by two brothers, William F.

and Henry W. Wendt. William, the older brother, established the company in 1878 to manufacture blacksmith's forges and later added a variety of blacksmithing equipment. Henry, the younger brother, was in charge of the manufacturing activities of the firm. He soon earned a reputation as an expert foundryman with a talent for successfully turning out difficult castings. His skill earned him a term as president of the American Foundrymen's Association. It was Henry Wendt who suggested that Buffalo Forge begin the manufacture of fans and heating equipment, and for some years he played multiple roles as the company's only salesman, designer, and manufacturer. His interest in this new field similarly led to his election as the first president of the National Association of Fan Manufacturers. It was the younger Wendt, then, who encouraged Carrier to pursue the more theoretical and pioneering aspects of heating and ventilating.[36]

Carrier worked primarily in the engineering department, designing heating and drying systems, but with the blessings of the Wendt brothers his work extended to the testing and rating of heating coils and fans. Thus, he soon alternated between design and research. His interest in humidity control began in 1902 when the Sackett-Wilhelms Lithographing Company of Brooklyn, New York, approached Buffalo Forge about controlling the humidity of their printing room. Fluctuating levels of humidity caused the paper to swell or shrink, effectively changing the size of the paper each time it passed through the press. The company did multicolor printing that called for several passes through the press, and it was plagued by the failure of the colors to register in the proper position. Sackett-Wilhelms appealed to Buffalo Forge for a ventilating system that would end its humidity-caused woes, and the problem was referred to Carrier.[37]

Working with Buffalo Forge engineer Edward T. Murphy, Carrier first tried to dry the air chemically with a calcium chloride solution, but ultimately he designed a fan-coil system similar to the one installed in the New York Stock Exchange. Buffalo Forge manufactured heating coils—tubes through which hot liquid was circulated—so it was natural for Carrier to try running cold water through the same coils to achieve a degree of cooling. He understood that cooling would provide the humidity control that the company needed for consistent production. Sackett-Wilhelms asked for a temperature of 70 degrees in the winter and 80 degrees in the summer, with 55 percent humidity year-round. Carrier's system was installed in 1902 and relieved conditions somewhat. However, like many engineers who came to atmospheric cooling from the heating profession, Carrier had underestimated the air volume necessary to effect cooling. The system lacked the capacity that it needed to solve Sackett-Wilhelms's problems completely, but it launched Carrier on a lifelong pursuit of atmospheric humidity control as part of ventilating systems.

Carrier continued to experiment with ways to produce precise levels of relative humidity. He rejected the fan-coil system and instead focused upon the air washer, a large enclosure fitted with a bank of spray nozzles and coupled with a fan that drew the air through the spray chamber. The air washer was first developed to wash dirt, smoke, and fibers out of the air supply. Carrier used it primarily to regulate the humidity level of the incoming air and only incidentally to clean it. Carrier reasoned that if he could saturate the air with a fine spray of water, at a specific temperature he would have a known quantity of moisture. The air as it left the spray chamber was always at 100 percent relative humidity, and it was then either heated or mixed with warm air to achieve a lower relative humidity for use in the factory. To humidify, the air was drawn through warm-water sprays, saturated, and then brought up to room temperature. To dehumidify the interior of the factory, Carrier pulled the air through an air washer equipped with chilled-water sprays to bring down the temperature; even though the air emerged fully saturated, at this low temperature it carried little moisture and produced relatively "dry" conditions once the air was raised to normal factory temperatures. The spray chamber of the air washer was crucial to Carrier's evolving ideas on the control of humidity levels. If the air left the conditioner fully saturated and was subsequently raised to a specified temperature, Carrier could control the relative humidity within the factory with great precision.

Carrier had first been drawn into humidity control by the dehumidification needs of a single company, but I. H. Hardeman, a fellow engineer at Buffalo Forge and a graduate of Georgia Institute of Technology's textile engineering program, immediately recognized the air conditioner's usefulness to the problem of increasing humidity levels in textile mills. When Hardeman became the Buffalo Forge representative in the South in 1906, he sold one of Carrier's air conditioners to the Chronicle Cotton Mill in Belmont, North Carolina.[38] Thus, Carrier joined Cramer in process air conditioning, the control of atmospheric conditions for better production.

Just when Carrier became aware of Cramer's parallel activities is not clear. But in late 1907, when Carrier proposed that the Wendts form a subsidiary company for heating, ventilating, and humidification, the company was named the Carrier Air Conditioning Company of America—adopting the term coined by Cramer nineteen months before. The subsidiary specialized in the

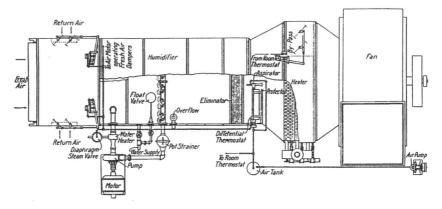

Carrier air washer with humidity control. Willis Carrier's earliest air-conditioning systems incorporated a spray chamber like this one. Carrier equipment treated the air in these massive central air-conditioning systems and distributed the treated air to the factory through ductwork. In this 1911 configuration of equipment, air moved from left to right through a mixing chamber, a spray chamber, a set of baffles, a heater, and a fan. Air from the outside and the factory was first drawn into the mixing chamber in varying proportions, then pulled through a set of water sprays at a specific temperature that saturated the air with moisture. The saturated air passed through a set of baffles (the eliminator) to remove any droplets of water. The saturated air contained a known quantity of water, which made it easy to heat the air so that the desired humidity was achieved before it was blown into the ductwork for distribution. A control instrument regulated the operation of the equipment. (ASME Transactions 33 [1911]: 1060)

design of air-conditioning systems and the manufacture of air washers. Carrier became vice president of the new company but served strictly as a manager with no ownership in the firm. He later speculated that the Wendts chose the name of the new venture as a way to avoid linking the Buffalo Forge reputation with a potential failure. The choice also served to disguise the fact that the subsidiary was competing with some of Buffalo Forge's steady customers who bought the company's fans for their own ventilation systems.

Willis Carrier became absorbed in the problems of humidity control, which for him constituted the heart of air conditioning. In 1911 he informed his engineering colleagues that "a specialized engineering field has recently developed, technically known as air conditioning, or the artificial regulation of atmospheric moisture."[39] A series of tests run at Buffalo allowed him to formulate what he called "rational psychrometric formulae," by which he hoped to determine more accurately the temperature-humidity relationship. Carrier did not consider the psychrometric data available from the U.S. Weather Bureau sufficiently precise for air-conditioning work. In a paper published by the American Society of Mechanical Engineers (ASME), Carrier proposed a formula for hammering out a more precise relationship between temperature and humidity consonant with theoretical principles.[40] While later modified by other researchers, Carrier's efforts attracted a good deal of attention for bringing scientific principles to bear on a very practical problem.

That same marriage of theory and practice, or science and engineering, was found in the companion paper, which described air-conditioning apparatus.[41] In this article, Carrier and fellow Buffalo Forge employee Frank L. Busey described the use of the air washer, formerly a device for cleaning the air supply, for the purposes of precise humidity control. Carrier's subsequent commitment to the technology of the spray chamber was part of his emphasis on humidity regulation as the heart of air conditioning and on precision as the key distinction between air conditioning and humidifiers. In the years when the Carrier companies dominated the competition, this mix of function, hardware, and standards of performance was irresistibly urged on the rest of the industry, and it soon represented the standard definition of air conditioning.

Despite Carrier's technical successes, Buffalo Forge became nervous about its subsidiary's commercial potential. In 1915, as war loomed in Europe, the senior partner at Buffalo Forge, William Wendt, began to worry about business stability. His uneasiness about business prospects in the wake of what he believed would be Germany's assured success was coupled with a conviction that Buffalo Forge should no longer compete with its customers by maintaining an engineering subsidiary.[42] The brothers decided to dissolve Carrier Air Conditioning and pull back to their core fan and forge enterprises. Although Carrier and Lyle would remain on Buffalo Forge's payroll, the entire staff of Carrier Air Conditioning would be fired. Faced with this crisis, seven air-conditioning specialists from Carrier Air Conditioning, headed by Willis Carrier and J. Irvine Lyle, established a new firm, Carrier Engineering Corporation (CEC) in 1915.

The survival of the Carrier firm in this new guise was pivotal as other pioneers became less active. Alfred Wolff had died in 1909. Stuart Cramer retired from business in 1918, selling his air-conditioning interests to the Parks Company of Massachusetts, an established company in textile humidification. By contrast, the new Carrier Engineering Company was exclusively concerned with air conditioning. Unlike both Carrier Air Conditioning of America and Cramer Air Conditioning Company, which were sidelines for their owners, the seven founding members of CEC were dependent upon the technology's success for their own survival. That wholehearted investment in the infant technology was important, and CEC became a prominent advocate for the new technology, supplying consumers and the public at large with a vision of air conditioning's powers and its potential.

That vision was more elaborate than those that either Wolff or Cramer had ever articulated. But as the industry developed, it became clear that air conditioning entailed a melding of ventilation and humidification into one system and the substitution of that mechanical system for window ventilation. From the beginning, Wolff, Cramer, and Carrier all worked to gain greater control over the indoor atmosphere by closing windows and drawing a sharp distinction between the inside and the outdoors.

CHAPTER TWO

Custom Production,
Industrial Processing,
and Engineering Guarantees

*A*ir conditioning gained the widest acceptance not among Wolff's urban elite seeking personal comfort but among the numerous factories that processed hygroscopic materials. Manufacturers were enthusiastic about the promised independence from natural climate. But the ideology of the "controlled" environment left unanswered the question of who was in control. Although technological change in the factory is often seen as a two-way conflict between capital and labor, a look at process air conditioning before 1929 suggests that there was instead a three-way struggle for control among engineers, managers, and workers.

The power of air-conditioning engineers in influencing factory development came from the dominance of custom production in the early years of the industry. The importance of engineering design for tailor-made systems gave designers considerable influence in their relations with both the company they worked for and the customer they designed for. As engineers designed air-conditioning systems for the factory, they quickly learned that to make good on their guarantees, they had to design for both the architecture of the factory and the activities within the building. In other words, they had to view the factory as both building and process. One consequence of this more sophisticated view was that engineers attempted to freeze both the building and the process within the original design parameters.

The continuing concern of air-conditioning engineers with factory processing brought them into conflict with both management and skilled workers. The challenge to their professional authority from both groups led them to adopt a quantitative and experimental approach to solving the problems of environmental control. This approach clearly shifted power to engineers in the struggle over who controlled the controlled environment.

It was the management of humidity added to traditional heating and ventilating techniques that convinced engineers they now had the skills to create artificial climate indoors. As it developed, process air conditioning for the factory was marketed with the promise that it would do more than mitigate unpleasant conditions: Many engineering firms guaranteed that they would provide specific atmospheric conditions. Increasingly, they began to describe air conditioning as climate control or man-made weather. Not only was the new air-conditioned factory precisely controlled, but it also provided a consistency that even the most perfect natural climate could not match. Air-conditioning advertisements generally promised independence from outside weather, and Carrier Engineering Corporation (CEC) in particular urged its clients to "make every day a *good* day."[1]

The conceptualization of air conditioning as artificial climate proved seductively attractive to certain groups of consumers who welcomed the escape from both climate and season that air conditioning promised to provide. That vision appealed most to manufacturers whose products changed their size, appearance, and handling characteristics as humidity levels within the factory fluctuated with outside conditions. In adverse weather cigarette machinery jammed, chocolates turned gray, and cotton threads broke. Textiles, tobacco, pasta, sausage, black powder, chewing gum, candy, chocolate, and flour were only a few of the industries affected by changes in humidity.

Air conditioning as artificial climate appealed to manufacturers in two ways: it promised to eliminate the effects of seasonal weather fluctuations and it suggested a way to defeat the effects of climate, making geographic location unimportant. One air-conditioning executive explained that "certain areas like Lancaster, in England, and New Bedford and Fall River, in America, were thought to be so favored by nature, no humidification of mills was at all necessary."[2] Noting the climatic advantages of the Northeast, a southern journal reiterated that "even though it was seldom ideal as now established for the highest quality product, it tended toward a more favorable result sufficient to determine the location of industry."[3] Man-made weather within the factory could eliminate the competitive advantage of those favored locations. Addressing the idea directly, one air-conditioning company promised in its 1912 catalog, "With our equipment the manufacturer is enabled to *defy climatic conditions* or weather fluctuations and to discount geographical location. We can give to a mill, *no matter where it is located*, the advantage of uniform atmospheric conditions not surpassed by the best textile localities of Great Britain."[4]

However, problems with humidity affected every region to some extent. Both temperature and humidity levels varied with the season almost everywhere. F. L. Becker, manager of the American Chicle plant in Long Island City, New York, noted that "the manufacture of chewing gum is affected to as great a degree, probably as any product, by the slightest change in weather conditions. Prior to the advent of the Air Conditioning System, the production of this article was dependent upon the whim of the weatherman."[5] Between 1916 and 1925 CEC did more than $1 million worth of business with manufacturers of chocolate, chewing gum, and hard candy.

By eliminating the effects of seasonal variation, manufacturers gained more control over the production process, assuring more consistent output and quality. For example, the Collins Baking Company in Buffalo, New York, installed an air-conditioning system in 1911 to maintain a constant temperature in its plant, thus controlling the rate of fermentation and holding production to a fixed schedule. The new system also regulated humidity levels to prevent both crusting and sliming, which resulted in sour dough.[6]

Yet the portrayal of air conditioning as artificial weather was more than a selling device designed to appeal to manufacturers at the mercy of climate and season. The idea had both an appeal and an impact within the technical community. The most striking result of characterizing the new technology as man-made weather was the practice of guaranteeing not the equipment at a rated capacity, but rather the environmental conditions the machinery would maintain. While it was relatively simple to guarantee the temperature and humidity of the air supply, it required a much more sophisticated calculation to guarantee atmospheric conditions within the factory.

To design a system properly, engineers needed to estimate the heat load accurately and install machinery of sufficient capacity to meet that demand. The amount of heat generated within a building often came from surprising sources. Engineers learned to pay close attention to the architecture of the building, the number of people occupying it, and the type of manufacturing taking place—factors that varied with each installation. Warren Webster and Company explained the individualistic nature of industrial guarantees to its customers. "General guarantees," they wrote, "covering cooling, which may be accomplished by different types of Webster Apparatus cannot be made in a satisfactory manner. Installations so equipped are usually designed to meet specific conditions and suitable guarantees will be furnished in each case upon request."[7] Each installation was unique, specially designed to fit a particular set of circumstances. Thus, CEC assured prospective customers in 1921 that "it is customary for engineers who specialize in the design of such equipments to treat each problem individually and develop the most efficient equipment for the specific requirements of the client."[8]

Such custom production is often associated with a bygone era of craft production, a historic relic superseded by mass production. That sense of the new pushing aside the old appeared as early as 1832, when Charles Babbage made a distinction between "making" and "manufacturing."[9] In his discussion of Henry Maudsley's work for the British Navy, he described a system of "making" that included the production of single-item goods with general-purpose tools to the specifications of the client at a relatively high cost. This description of "making," or custom production, and the language to describe it prevailed for nearly a century. In 1914 Frederick Halsey, long-time editor of *American Machinist,* self-consciously used the same terms in his machine shop textbook, to distinguish between the work of the shop and the work of the factory.[10]

When they sat down to write their own book on shop practice in 1935, Joseph Wickham Roe and Charles W. Lytle borrowed heavily from Halsey and reproduced the "making" versus "manufacturing" distinction, but this time with a difference.[11] Substituting the term *building* for *making,* their characterization of custom production focused less on the methods of production and more on the distinctive economic and social relationships that flowed from its practice. "The two processes of production," Roe and Lytle wrote, "from initial sale to final acceptance follow different courses."[12] Custom design depended upon a direct relationship between buyer and seller, devolving upon the engineer an important role from beginning to end. In this system of production, according to Roe and Lytle, "the sale precedes the building and even much of the designing, and the engineer is intimately concerned in the selling as he must convince the purchaser of the superiority of his design."[13] Their description of custom production in all its commercial trappings vividly depicts the features of a vital system of production rather than the shadowy outline of a dying craft.

Roe and Lytle's assessment is useful not simply for their sensitivity to the economic aspects of custom production, but even more for their reportage of a modern variant of the building tradition. They conceded that "building methods will always have their place and are the only ones possible for a large and unstandardized work which must be made to suit special conditions. Great progress has been made, however, particularly in America, in the partial standardization of such work by standardizing the subassemblies and other details employed."[14] They cited the machinery for materials handling as the perfect example of standardized parts adapted to individual needs. The combination of the two methods—custom design with manufactured subassemblies—drew their approbation, and they pointed out that it "may often be the wisest and most profitable method of production."[15] This distinctive style might be characterized as a combination of the two classic systems of making and manufacturing, but it is more accurately defined as a variant of the former, since the economic relationships and commercial realities of this method of production differ little from the classic system of "making" or "building." From the beginning, air conditioning conformed to Roe and Lytle's pattern of building custom-designed systems out of standardized parts.

Many custom-designed goods are site-specific—that is, they are unique to the extent that they are adapted to the geography of the place they occupy. Most civil engineering falls into this category, as does architectural design at its best. In contrast, some scientific instruments and machine tools are produced to the particularities of clients' needs rather than the demands of location. Yet whether adapted to specific location or special need, custom-designed products are sold primarily on the basis of their qualities or performance characteristics. This particularity inevitably requires the talents of an engineer or technologist. Custom production, with its attendant economic relationships, places primary emphasis upon the engineer as designer.

Industrial air-conditioning systems were all tailor-made. They were designed by engineers, and the parts of the installation were furnished by different manufacturers. For example, in September 1911 consulting engineer Edward H. Vitaluis of Detroit sold to a bakery an air-conditioning system that consisted of American Blower fans, Carrier dehumidifiers, and York Manufacturing refrigeration machinery. Such installations were designed specifically for each factory building and for a particular industrial process. They were both site-specific and need-sensitive.

The importance of engineering expertise was heightened by the industry's practice of offering engineering guarantees rather than machinery guarantees. A Carrier engineer recalled the power of that kind of guarantee to clinch a business deal. A company representative reported that he faced a lower bid from rival B. F. Sturtevant, and persuaded the client to read the Sturtevant proposition aloud so that the two could be compared word for word. As he remembered, the exercise revealed that "both quoted upon the same basic essentials, that is, air washer, heater, fan, ducts, and the controls. The situation got worse and worse until they finally came to the guarantees and that let the cat out of the bag. Sturtevant had guaranteed the volume of air which the system outlets would handle and that all of it would be delivered from the outlets at a relative humidity of 70%; Carrier guaranteed that the system would maintain 70% in the mill."[16]

Guaranteeing atmospheric conditions throughout the mill rather than at the outlet was a strong selling point, but a more difficult engineering proposition. Atmospheric conditions were the product of a dynamic relationship between the air-conditioning system and factory processes. Designing a system that could respond to changing conditions within the factory required the sort of precise estimate of heat generated by lights, people, machinery, and industrial processes that Carrier became adept at providing. These kinds of guarantees reinforced both the need for custom design of systems and the primacy of engineering expertise.

Atmospheric guarantees changed the way engineers estimated the equipment necessary for any given job. Older methods of estimating the required humidifying capacity were based upon the floor space or cubic content of the mill. One engineer called this rule-of-thumb method "futile and foolish," for the requirements necessary to counteract the weather alone were substantial. A mill located in the southern lowlands required 30 percent more evaporative capacity than it would if located in New England. The type of construction of the building and the kinds of production that took place also made a difference in the performance of the air-conditioning system. Noting that horsepower inevitably became translated into heat, one observer calculated that a room with a low ceiling organized for spinning duck fabric for tires required nearly double the evaporative capacity per unit of content needed by a room in a mill of more modern construction engaged in spinning hosiery yarns.[17]

In his first textile installation, Carrier learned the importance of matching the capacity of an air-conditioning system to the activities within a building and not just to the building itself. In 1906 Carrier designed a system for the Chronicle Cotton Mills in Belmont, North Carolina, without visiting the site. His inspection of the completed installation was his first visit to a textile mill and a revelation of the working conditions within such a factory. He immediately recognized that he had failed to calculate correctly the capacity of the air-conditioning machinery. Carrier recalled, "When I saw five thousand spindles spinning so fast and getting so hot that they'd cause a bad burn when touched several minutes after shutdown, I realized our humidifier was too small for the job. All of the heat played havoc with relative humidity—raised air temperatures far beyond what we had calculated."[18]

This awareness of the role of heat generated within the factory was new to the humidification industry. Traditionally, companies that handled industrial humidification had made design estimates on the basis of building size. *Kent's Mechanical Engineers' Pocket Book* (1895), which one engineer called "the Bible of heating engineers," advocated a rule of thumb based upon total volume of the building.[19] As one manufacturer of humidifiers remembered common practice in 1906, "Textile manufacturers had been content to buy—and humidifier manufacturers had been content to sell humidifiers on a 'perhaps' basis. So many humidifiers to about so many cubic feet."[20] Heating and ventilating engineers had refined that formula from the simple calculation of the cubic contents of a room to a method for considering the exposure of the walls and the role of outside climate. Yet even the imporved methods for estimation produced occasional but spectacular failures. With a new awareness of the role of heat generated within the factory, Willis Carrier began to estimate 2,520 BTU per hour for each horsepower used to drive the machinery, while Stuart Cramer used the value of 2,550 BTU per hour.[21] Another air-conditioning pioneer called this value "one of the real keys to the calculation of the size of equipment."[22]

Air-conditioning companies could meet their atmospheric guarantees only if the designer was familiar with the requirements of specific industrial processes and designed the system accordingly. The engineers thus had to acquire industry-specific knowledge in order to gauge the volume of air or refrigeration capacity required. New York consulting engineer Walter Fleisher believed that the need for industry-specific engineering expertise was one of the factors that kept the number of air-conditioning companies small. He explained, "To cope with industrial air conditioning, one had to be familiar with the industry and as the need for an air conditioning installation was sporadic in any one vicinity only those companies who were nationally involved were able to maintain an air conditioning engineering force able to cope with the widely spread requirements for installations or engineering works."[23] This reliance on expertise led to the dominance of the industry by a small coterie of air-conditioning companies.

The importance of engineering expertise does not mean that patents played no role in the industry. Patenting activity in this period centered on control instruments for regulating temperature and humidity. There were at least three significant groups of patents among the pioneers. Cramer patented a regain control device in 1904, Carrier patented a dew-point control in 1906, and William G. R. Braemer patented a similar one in 1907.[24] In 1936, when Carrier recalled the pioneering air-conditioning companies, he named S. W. Cramer Company, Carrier Air Conditioning Company of America, and Warren Webster and Company—companies that installed systems based on these three patents.[25] This taxonomy of the early industry reveals Carrier's own preference for defining it in terms of inventions and patents. As he was the holder of one of the three pivotal patents, this organizing principle ensured his own fame as a pioneer, especially because some regarded the patented control device of his one-time coworker Braemer as essentially derivative of Carrier's own invention despite its separate patent rights.

In addition to these patent holders, at least three instrument companies manufactured temperature and humidity controls: Taylor Instrument Company, Powers Regulator Company, and Johnson Service Company. As early as 1902 Alfred Wolff installed a Johnson Service Company humidistat in the Andrew Carnegie residence for the automatic regulation of humidity, and from 1906 on American Blower Corporation apparently did all its air conditioning with controls from these three companies.[26] It seems clear that control instruments were one element in the competition of the new industry, but not the decisive one.

In contrast to Carrier's list of industry leaders, Fleisher estimated that there were probably only three air-conditioning companies active on the national level in the early years: Carrier Engineering Corporation, American Blower Corporation, and his own firm, W. L. Fleisher and Company. Each of these companies rose to leadership through engineering skill.

Indeed, the convergence of custom design and atmospheric guarantees resulted in an oligopoly of engineering firms. This made it easier for the industry leaders to control the information essential for success. Estimating heat loads accurately and sizing the equipment accordingly through an acquaintance with the industry and a knowledge of engineering principles were essential skills for meeting atmospheric guarantees. Among the ventilation experts in 1906, such information was valuable and not widely known. At a meeting of heating and ventilating engineers, Edward Berry of Philadelphia broached the subject of rules for estimating the size of equipment. "I think there must be some of the older men of the Society who have adopted rules which are accurate enough." But, he lamented, "there seems to be no way of getting track of such trade secrets."[27]

The commercial importance of such data undoubtedly encouraged secrecy. Monte Calvert has argued for the nineteenth century that among machine shops producing specialized machines for industrial customers, commercial relationships were grounded in personal relationships and trust, not innovative techniques. As a consequence, technical information flowed freely from one shop to the next.[28] The air-conditioning industry did not conform to that pattern. Unlike the clients of the earlier machine shops, the majority of air-conditioning customers were buying their first system and had no prior experience with the firm with which they contracted. In this new industry, technical competence was not only central to fulfilling guarantees, but also was the key factor in commercial competitiveness. Not surprisingly, any technical advantage that could not be protected by a patent was maintained as proprietary knowledge. As engineer Phillip L. Davidson recalled, "All engineering knowledge was very closely guarded as trade secrets, even such first grade principles as the fact that a machine requiring 1 h.p. to operate would liberate 42.5 Btu. per minute."[29]

Such principles of engineering design have been identified by Edwin Layton as engineering science, and he has persuasively argued for the role of engineering science in establishing engineering as an independent "mirror image twin" of the scientific community.[30] Yet little has been said of the role of engineering science in the commercial realm. Engineering knowledge can be regarded as basic engineering science or as proprietary knowledge. The engineering community saw it both ways.

Those who thought in terms of engineering science imagined professionalism as the answer to trade secrecy. In his 1917 presidential address before the American Society of Heating and Ventilating Engineers (ASH&VE), Harry M. Hart predicted that engineers would turn to the society to gain that expertise from other engineers. He linked the "increasing demand for suitable equipment to meet new requirements in heating and air conditioning" to an increase in membership. He believed that these new members were motivated by "the need of a closer co-operation among engineers in order to be better able to meet the new demands."[31] The ideal of professional cooperation had suffered, however, when the society organized a session on unusual air-conditioning and drying installations in factories but then experienced difficulty persuading engineers from active companies to contribute papers. "A seeming reluctance by some of our members to present papers of this character," Hart scolded, "would indicate a lack of accurate data or a mistaken idea that the giving out of such data would result in the loss of some advantage, presumably over competitors."[32] He continued, "This subtle reluctance that seems to grip so many of our members has been one of the greatest drawbacks to the advancement of the science of heating and ventilating."[33]

Within the air-conditioning industry, manufacturers and engineering firms were divided on the subject of promoting the free flow of technical information. Generally, manufacturers of component parts, such as fans, air washers, and refrigeration equipment, readily disseminated information that would help their customers use their products most successfully. Like early automobiles, air-conditioning equipment was generally assembled from the parts supplied by a variety of manufacturers. It used standard components common in the ventilation industry, even though each system was assembled according to the needs of the client. Manufacturers' catalogs often contained "textbook" sections that included simple tabular data as well as discussion of the basic engineering principles behind the problems customers faced. One engineer declared that "the manufacturer of heating apparatus is the greatest disseminator of knowledge concerning apparatus there is," while another marveled that any blower manufacturer "will send any one any information that he may ask, even to the extent of a whole library."[34]

The Buffalo Forge Company was one such manufacturer. Its catalog noted, "The Buffalo Forge Company has always taken the stand that engineering data and developments should not be hoarded as hidden treasures but should be made available for the use and edification of the engineering profession in general."[35] Willis Carrier's psychrometric table was one example. Although he used the U.S. Department of Agriculture psychrometric tables published in 1900, they did not provide the precision that he needed for air conditioning. With support from Buffalo Forge, he derived his own values and published the resulting table in the textbook section of a 1906 company catalog.

Engineering companies, on the other hand, were less likely than manufacturers to reveal the engineering data behind their successful installations. Carrier personally made the transition from working for a manufacturing company to running an engineering firm with a consequent change of perspective. From 1907 to 1914, he worked at the Buffalo Forge Company subsidiary, Carrier Air Conditioning Company of America. He cherished the ambition, conceived in 1905, of publishing a handbook on air conditioning, and his professional papers were part of that greater plan. As his biographer described it, "when no paper was read, and therefore none published, on a subject he wanted to use as a reference in his handbook, he or [Frank L.] Busey would write the paper, present it, and when it was published he'd have his reference."[36] The handbook, *Fan Engineering*, appeared in 1914 while Carrier was still at Buffalo Forge.

At Carrier Engineering Corporation, Carrier's position was distinctly different. CEC was clearly an engineering firm, in which expertise was the principal commodity. With the exception of control instruments, they did no manufacturing, instead buying their fans from Buffalo Forge and their refrigeration equipment from York Manufacturing Company.[37] The creation of air-conditioning systems out of standard parts supplied by a variety of manufacturers was essentially the same process in both the old and new company, but a patent agreement between Willis Carrier and Buffalo Forge, which gave the latter half the rights to all of Carrier's patents, extended beyond the term of his employment with the company. Until the expiration of the patent agreement, Carrier could not compete on the basis of innovative hardware without yielding half the advantage to his old company. Thus, CEC began life as an engineering firm. As the company told its customers, "We are not interested primarily in the manufacture and sale of equipment for this is an organization of engineering specialists."[38]

As an engineering firm, the new company pursued an explicit policy of maintaining a measure of secrecy about its methods. "For many years," recalled L. L. Lewis, a CEC engineer, "we refused to divulge the air-handling capacity of the system and revealed only its requirements for floor space, power, water, and steam."[39] In particular, the company was not forthcoming about the details of how it estimated designs. Lewis called CEC's competitors "troublesome rather than effective," noting that "no consulting engineer had then developed the skill that would enable him to specify air conditioning."[40] Carrier's 1906 insight at Belmont—that machinery, people, lights, and processes all produced a substantial amount of heat that affected air-conditioning performance—turned the company's attention to the mill interior rather than its exterior. Carrier's ability to attach a quantitative figure to many of the activities that occurred within the factory gave the new company a competitive edge. One company member recalled that "up to about 1925 Carrier people had, to all practical purposes, a monopoly on the brains and the know-how of air conditioning. Consequently its policy was to educate its own people but at the same time take extreme measures to prevent this knowledge and practical experience from getting into the hands of outsiders—especially competitors but including consulting engineers and contractors in pretty much the same class."[41] In this new circumstance, CEC's British subsidiary now complained that Carrier's *Fan Engineering* was too informative, and the company agreed to refrain from distributing it in Britain.[42]

Indicative of this concern for control of engineering information was the fate of the company newsletter, *Carrier*. Circulated within the company, it featured accounts of recent contracts, triumphs, and problems. The masthead bore the statement, "The *Carrier* is mailed personally to each member of the Organization, with the understanding that it is to be regarded as entirely confidential and for the information of the original recipient only." Lewis remembered that the company's concern about the information in the *Carrier* extended beyond this caution. "Very few copies are now in existence," he wrote. "This is the result of a statement in a subsequent issue to the effect that the right to receive carried the responsibility to destroy."[43]

The importance of industry-specific knowledge in the air-conditioning industry was a variation on the common experience of consulting engineers of all kinds. Knowledge about industry practice was often something that engineers picked up in the factory. Like other consulting engineers, air-conditioning engineers knew their clients were concerned that knowledge gained in one factory not be used to benefit that manufacturer's business competitors. A CEC pamphlet stressed that work was done "with the agreement that the information obtained should be kept strictly confidential."[44] An example of this phenomenon was the company's involvement in the linoleum industry. CEC set up an air-conditioned drying system for a linoleum manufacturer, but the customer had difficulty with "stickers," tiles that swayed in the draft of the dryer and stuck to nearby tiles before they were dry. Although its engineers successfully solved this problem, CEC executives wrote its overseas subsidiary that "we were under an agreement with the Armstrong Cork Company, Linoleum Division, whereby we could not place at your service the information gained at the Plant in the curing of linoleum."[45] The position of the early air-conditioning companies as consulting engineering firms reinforced the tendency toward maintaining engineering expertise and knowledge gained on the job as secret, proprietary knowledge.

Because air conditioning was still a developing industry, an engineer's "experience" quickly became "expertise." Each installation presented new problems that company engineers solved on the job, and those solutions informed subsequent practice. The continuing importance of practical experience was made clear by John R. Allen, professor of mechanical engineering at the University of Michigan, in the preface to his 1905 textbook on ventilation. He noted that "the design of heating and ventilating systems has not been reduced to an exact science . . . One reason for this is the lack of exact experimental data governing some of the most important factors entering into these calculations. This lack must be filled from the designer's experience."[46] Since a separate research department was not established at CEC until 1919, the factory itself became a laboratory for the company's engineers, and industrial applications became the main avenue for advancements in the art. But this phenomenon posed a dilemma for the company: how could a firm's competitiveness be based on highly individualized engineering skills? CEC's answer was to convert individual engineering skill into corporate expertise. In early 1916, each engineer recorded his experience and deposited it in company files which were called "Confidential Data" files.[47] This archive formed the heart of CEC's "unpatentable" engineering knowledge. Organized by industry, the reports were authored by the company's specialist in each field and represented a summation of his experience.

The extent to which the firm could effectively expropriate individual skill was problematic. In a new field in which so much was unknown, the Confidential Data files record a style of engineering that was often highly intuitive. The "art" of air conditioning, which depended upon an individual's intuition informed by experience, proved difficult to reduce to paper. One engineer wrote frankly in a document deposited in the files: "There are so many independent variables in this part of the proposition that it is impossible to carry the calculations in logical order and proper sequence. The writer had a fairly accurate knowledge of many of the sizes from previous experience, and consequently was able to assume certain factors quite accurately. In the case of others, a method of cut and try, which is not indicated here, was followed."[48]

Engineers themselves referred to air conditioning as "the art of ventilation." They sometimes used the phrase in an older usage to mean simply the practice of ventilation. By the turn of the century, however, some engineers insisted upon using the new term, "science of ventilation." It became clear that the older language expressed a side of engineering not encompassed by science. The "art" included the solution of engineering problems where clearly defined mathematical relationships, constants, or measurable factors were not present to dictate a sure course of action. It emphasized creativity over knowledge. It often meant engineering solutions based on intuition born of experience. This contrast between the art and the science of engineering mirrored the conflict between shop culture and school culture in the larger realm of mechanical engineering. Clearly, business displayed a greater affinity for "science" than for "art," empowering the corporation rather than the individual. It was the unambiguous science of engineering that was most easily captured and broadly disseminated within the company, independent of an individual engineer's personal participation. Despite the difficulties of recording individual expertise, however, the Confidential Data files proved a valuable resource for CEC.

As air-conditioning engineers increasingly became concerned with controlling the production process, they began to argue the benefits of air conditioning in controlling the work force as well. Air-conditioning companies extended their predictions to workers as well as production to appeal to manufacturers. Clearly, the concept of a totally controlled environment lurked behind the idea of man-made weather. But besides the intrinsic appeal of that vision to engineers, it was articulated just when the technical community had begun to concern itself with issues of labor, machinery, and workplace control. Thus at the same time that Willis Carrier presented his psychrometric formula to the American Society of Mechanical Engineers (ASME) as the basis for an understanding of temperature and humidity, Frederick W. Taylor was expounding the science of efficiency. And as Taylor had claimed for his new management strategies, air-conditioning companies portrayed the new technology as an advance that benefited all members of the industrial community. However, the adoption of air conditioning reordered the balance of power between the different parties that composed that community. What is striking about the consequent realignment of power in the factory is the extent to which engineers increased their authority relative to both management and workers.

The conceptualization of air conditioning as manufactured weather expressed not only the potential of the new technology in a way that might appeal to manufacturers but also the brash self-confidence of the air-conditioning engineers. The implementation of this mechanical ideal relied upon their knowledge and skills, but it also depended upon the willingness of manufacturers and workers to yield atmospheric control of the factory to a technical elite. Like atmospheric guarantees, the concept of artificial climate placed the emphasis on air conditions rather than machinery. Similarly, its successful realization was dependent upon correctly judging the dynamic relationship among the machinery, the building, and the occupants. And in ways that, perhaps, were not quite anticipated, the successful installation of any system of artificial climate required that the air-conditioning company control all three elements. Thus, engineers attempted to fulfill both their guarantees and the promise of the technology by sealing windows, closing doors, and freezing work patterns and production processes. To achieve environmental control, air-conditioning engineers insisted on control, first, over the factory building and then, increasingly, over the activities within it.

Maintaining atmospheric control over the building led to a persistent involvement in production processes themselves. A larger role for air-conditioning engineers in the work of the factory challenged and circumscribed both the decisions of management and the prerogatives of workers. In fact, clients that operated either the air-conditioning machinery or their production processes in unanticipated patterns were a constant irritant to an engineering company hostage to its atmospheric guarantees. The conflict over changing patterns of production is illustrated by CEC's exchange with Atlas Powder Company's Wolf Lake plant.

Wolf Lake was using a CEC installation to dry black powder when the company began to complain to CEC about excessive moisture in the final product.[49] Faced with a balky system, Wolf Lake personnel "attempted to help themselves out of their trouble by several methods of baffling," but CEC engineers concluded grumpily "that some of the baffles are undoubtedly interfering with proper circulation in the room."[50] CEC engineer H. B. Forbes was sent to investigate, perhaps aware of his boss's assessment that "it will be helpful to all of us if we can sell the local people on the idea of this dryer. They are apparently a bit old-fashioned and this type of dryer was purchased over their objections."[51] Forbes had to wait to get half a dryerful of powder due to an explosion a few days before his arrival, but his ultimate conclusion was that the company had been using a batch dryer as a continuous dryer, adding a new load of wet powder before removing the first load.[52] CEC insisted upon batch drying even though it might slow production. Atmospheric guarantees meant that the company had a continuing interest in how its product was used and some little power to reorder the factory around the needs of the air-conditioning system.

As the Wolf Lake example shows, CEC engineers, sent to resolve complaints by clients about system performance, were likely to complain in turn about the independent actions of factory officials. Factory personnel closed dampers, built baffles, and ran the machinery contrary to design assumptions.[53] A year later the Wolf Lake superintendent was alarmed to find black powder in the return air duct and began running the machines on all outside air, even though the system had been designed to handle a given percentage of recirculated air.[54]

44

Efforts to make good on guarantees often pitted engineers against their clients, with engineers insisting on a larger role in factory production methods or schedules.

Indeed, the most comfortable position for the air-conditioning company trying to make good on an atmospheric guarantee was to freeze the production process around an industry standard. If, for instance, all rayon plants adopted the same processes, they would present the same engineering problems for the air-conditioning engineer. CEC counted on some standardization within the industry to make the heavy initial investment in engineering work pay off in subsequent installations. For that reason CEC declined to take a job for a company that dried golf clubs because it felt the potential number of installations was too small to warrant the time involved. In general, diversity in production methods created uncertainty in new installations; improvements in processing overset design parameters. In one instance, CEC engineer A. E. Stacey noted with exasperation that the problem was not with the company's air-conditioning system but with the client, in this case Amoskeag Manufacturing Company. Amoskeag had a "lack of knowledge of the business," Stacey maintained.[55] Using his general experience in the industry to suggest that Amoskeag was running one process too long, he charged that the company "did not know how to make rayon."[56] We do not know what Amoskeag officials might have thought had they known Stacy's opinions, but it is clear that such assumptions inevitably put air-conditioning engineers on a collision course with management.

No less of a conflict evolved between air-conditioning companies and factory workers. For workers in hygroscopic industries, an important part of their skill involved an understanding of the complexity of natural material and its response both to weather and to mechanized processing. Manufacturers that processed hygroscopic materials watched the weather carefully. If humidity levels dropped, cotton thread would become brittle and break; if the humidity rose, chocolate would develop a gray, powdery coating. A surprising variety of enterprises were affected when unchanging machine processes encountered variable natural materials. Weather-sensitive materials strongly linked the factory to the natural world—a world that air-conditioning engineers hoped to replace with manufactured weather.

In the textile industry, for example, spinners and weavers regulated windows to change environmental conditions. While managers generally operated the mechanical equipment, windows were within easy reach of most workers. Windows therefore became a contested ground, for managers were never sure that workers were trying to achieve the best production conditions. While closing windows could increase humidity within certain limits, that increase was almost always achieved at the expense of ventilation and personal comfort. Because of the conflict between the needs of production and the conditions of comfort, the manner in which workers chose to regulate the windows under their immediate control was not easy to predict. Those who were paid by piecework rates might decide to keep the windows closed in warm weather to preserve high humidity levels and increase their productivity despite the personal discomfort, or they might sacrifice wages for better working conditions.

That conflict between process and comfort undermined the authority of factory operatives who were often experienced judges of proper environmental conditions in an inexact craft. The proper levels of humidification varied with the type and size of the yarn or cloth being produced, the length of time the fibers were exposed to the conditioned air of the workroom, and the type of mechanical processes being used, among other variables. CEC's textile expert, E. P. Heckel, admitted in the Confidential Data files that it would be "quite a proposition" to itemize the different humidities required for the different departments and different grades of cloth. Not only did variety make this task difficult, so did difference of opinion. "I have known some spinners that would be delighted if you maintained for them relative humidities of 50%," he wrote, "and other spinners who would jump at you with all fours if you tried to give them less than 65%."[57] CEC routinely supplied 60 percent relative humidity for cotton spinning and 75 percent for cotton weaving.[58] Nevertheless, it is clear that although these humidity levels were produced with great precision and consistency, the standards themselves were averages; they did not represent any greater insight into the response of natural materials to weather fluctuations than that supplied by experienced spinners, weavers, or overseers. What an air-conditioning system did supply was greater capacity, automatic controls, and an end to conflict between labor and management over the manner in which windows were regulated.

Despite the considerable knowledge that factory workers possessed about environmental conditions, engineers insisted upon total control over the environment. Artificial climate displaced natural climate and the skills that went along with it. A series of air-conditioning advertisements by the Parks-Cramer Company makes the underlying issues quite graphic. A series of eleven advertisements drove home the point that open windows in the textile mill were bad for production and profits.[59] One proclaimed that "spinners can't be trusted with a window"; another advised owners to break the "window-lust of your spinners."[60] The text that followed these provocative statements impugned not the motivation of workers but their skill at atmospheric control. "Asking spinners to create uniform conditions in the spinning and weaving rooms by adjusting windows is like asking untrained weather observers to give an accurate weather report," the ad maintained.[61] While the cut-out doll figures of female spinners and male supervisors in the advertisements seem to depict the conflict between labor and management, the unseen figure of the engineer is really the chief actor in this drama, for the struggle is, in essence, one between craft knowledge and engineering science.

The rivalry between CEC air-conditioning engineers and skilled workers can be seen more clearly in other industries. Maltsters, the men who supervised the sprouting of grains for breweries, considered control over atmospheric conditions part of their job skills. At the Kirkpatrick Malting Construction Company in 1914, the maltster refused to have anything to do with a new air-conditioning system and left the engineer from the air-conditioning company to handle the first test of the system. After a successful result, the maltster confided, "I gave you barley for the try-over which was so inferior that I would not have used it. Still you get fine results. The system is good all right."[62]

Perhaps the best documented example of the impact of the new technology on labor can be found in the macaroni industry. Macaroni factories displayed a pattern of intermittent production typical of industries that processed hygroscopic materials. As one journal explained, "macaroni cannot be made every day in the year or month or week. It must be a dry day, or the substance from which it is made will not bind properly."[63] Particularly difficult was the drying of pastes into the finished pasta, a complex operation that depended upon the product's shape, the factory conditions, and the weather. Excessive humidity promoted the souring or spoiling of macaroni through bacterial growth, while excessive heat dried the product only on the top layer, sealing in the moisture below and leading to checking, cracking, and breaking. As one trade journal noted, "the length of time required to cure or dry macaroni and spaghetti varies according to the process employed and how much the process is affected by the atmospheric conditions outside the factory and the standards of quality maintained. Some manufacturers complete the process in three days; others require five or six days."[64] Successful drying of macaroni depended in part upon the skill of the workers. A report on the numerous Philadelphia manufacturers singled out one: "The factory of Frank Cuneo is scrupulously clean and the workmen employed unusually skilled at their trade. They have the best luck in successfully 'curing' the large quantity of pastes made by the factory."[65] By

1907, however, one Bowery manufacturer confided to the journal's correspondent: "So far as making macaroni is concerned, it seems to me there is room, in the way of machinery for decided progress and improvement."[66]

CEC soon preceived a new market for its skills: applying humidity control to the problem of industrial drying. After initial tests in the Bellanca Macaroni Company's Buffalo plant in 1916, CEC signed contracts with three macaroni manufacturers, Skinner Macaroni Company, Crescent Macaroni and Cracker Company, and Foulds Milling Company in 1917–18.[67] It is clear that engineers saw the advantage to manufacturers of replacing skilled workers with those who were less expensive, less troublesome, and less knowledgeable. The Confidential Data files record how CEC engineers viewed the mechanization of macaroni drying, and perhaps also the basis on which the installations were sold to manufacturers. CEC macaroni specialist Russell Tree wrote:

> Some manufacturers have an idea that only the Italian understand the drying of macaroni. That may have been true at one time, but today it is absolute rot. When I say it does not require macaroni skill, I mean that it does not require a man familiar with macaroni to dry it with our dryer. All a man needs is a regular amount of human intelligence. You may wonder what that may mean to the manufacturer. It means that instead of a small field of men to hire from, he can chose [*sic*] from a large field. Then too, if a man quits, there is no great loss, for a new one can be taught in less than a week to do with our dryer what in the past men have taken years to learn.[68]

Quite plainly, this was technology in the service of management. For the manufacturer, part of the appeal of the new technology was to break the hold of skilled workers on production schedules and techniques, and give this control to management.

Yet the technology that CEC developed did more than replace skilled workers. In a move that set him in direct rivalry with workers, Carrier believed that he could improve upon traditional drying techniques, not simply duplicate them. Rather than give these macaroni companies atmospheric guarantees, the company was emboldened to guarantee production rates. Preliminary tests at the Bellanca factory suggested that drying rates could be gradually increased rather than held constant over the drying period, and he installed in the Skinner plant a new control instrument, a thermotyne invented by H. Y. Norwood of Taylor Instrument Company, which gradually lowered the dewpoint setting with a cam.[69] Carrier predicted, and CEC guaranteed, a twenty-hour drying time.

The results were disastrous. Before the macaroni was completely dry, it cracked and fell off the drying racks onto the floor. CEC engineers Tree and R. Leslie Jones filled the drying racks again and within twelve hours another 10,000 pounds of macaroni had collapsed onto the floor. CEC had no formal research department until 1919, so Carrier, Tree, Jones, and a Taylor Instrument representative turned the Skinner plant into an experimental laboratory. Tree recalled, "We ruined a lot of macaroni before arriving at the minimum and safe drying period, and we paid for the ruined macaroni at the rate of five cents a pound. We walked miles during the tests, from one drying room to another. A man from Taylor Instruments companies (who inspected controls) ended his first day in his sock feet."[70]

None of the macaroni installations met its guarantees. The customer refused to finish payment on the system; two years later it sued for the return of the initial payment of $13,936. The courts decided against Skinner, but in retaliation—whether by management or by workers is unknown—CEC's erecting man, R. B. Winfrey, was beaten up by two unidentified assailants.[71] Tree evaluated the experience for the company's Confidential Data files: "We have never met with any serious difficulty in producing real high grade macaroni; all of our drying plants have done this but our troubles have been due to the time limits of our—let's be frank—foolish guarantees."[72]

The failure of CEC to replace traditional craft skills with mechanical systems appears to be largely due to its failure to recognize the complexity of the relationship between natural materials and environmental conditions. Marveling over that complexity, one engineer wrote that "the commercial problem with respect to drying hinges upon the fact that no two macaronis have exactly the same drying characteristics."[73]

The macaroni fiasco was not the last time that the company underestimated the complexity of natural materials. In 1919 CEC formed Tobacco Treating Inc., and large-scale experiments were begun in drying tobacco leaves. The most prized leaves were those that emerged from drying with a rich golden color and with excellent handling qualities; these were used as the "wrappers" for cigars and brought a good price. CEC's drying process, achieved through mechanical environmental controls, yielded leaves all of which had that rich gold color. Yet further experiments revealed that these machine-dried leaves had none of the superior working qualities of high-grade tobacco produced under craft methods. The president of Porto Rican–American Tobacco Company advised CEC that "the cigar makers have refused to work anymore of the wrappers cured through your process on account of the same being entirely too tender."[74] To Carrier he explained, "The cigar-makers and strippers refuse to work it on the ground that it breaks in their hands."[75] Not only had engineers failed to produce wrappers with good handling characteristics, but also the new curing process eliminated the color difference that had allowed cigar makers to distinguish between fair- and good-quality leaves. The tobacco treating enterprise collapsed.

In both of these cases, engineers had before them a model of the process of drying. Had they been interested simply in aiding management's attempt to break the monopoly of skilled workers by duplicating their skills, that model would have been the easiest to follow. But CEC engineers chose to try to radically improve traditional methods, greatly accelerating the process or dramatically changing the product. They pursued a different path, and from it we can extrapolate a different goal. This drive for dramatic technical change was fueled by the recognition that skilled workers were the engineers' greatest rival to technical expertise in the factory; a substantial change in technology would put engineers unquestionably in control.[76]

In both macaroni and tobacco drying, CEC had not been able to improve upon the product of traditional craft practice. Engineers clearly knew less about the complexities of natural materials than skilled workers did. Part of the difficulty that they experienced was a clash between the quantitative approach of engineering and the sensual approach of craft production. In yet another example of this, Walter Fleisher lamented his own failure in drying. He had calculated that to dry skins for a leather company, his equipment should be able to remove 10 percent of the water by weight. Yet even after the removal of 17–26 percent, the client claimed the skins were still not dry. Only the removal of an additional 1.5 percent satisfied the client. "Their method of testing a skin to see whether it was dry," Fleisher recounted, "was entirely by feeling. They seemed to know what they considered was a dry skin, and cared very little for percentages of moisture removed."[77] Once again, engineering approaches to the processing of natural materials failed to replicate older methods, and craftsmen were often the judges who set the standards. In the eyes of these workmen at least, quantification led to oversimplification and not to precision.

Carrier's response to Fleisher's dilemma and his own was to call for more science, not less. He argued that the appropriate level of residual moisture in a properly dried material could be experimentally determined and quantified. He called upon scientific laboratories to produce such data for a range of materials and for engineers thus to arm themselves.[78] The air-conditioning engineer's dilemma was that he was in direct rivalry for control of production processes with skilled workers whose claim to authority was experiential knowledge. Laboratory constants and quantitative standards were essential advantages to engineers who had lost the first round in the match between craft and engineering.

In a way that was consonant with the popular enthusiasm for "science" and "efficiency," air-conditioning engineers thus fell back on science and quantification as a way to bolster the uncertain performance of environmental engineering in the factory. For a large body of workmen at the turn of the century—especially those employed in hygroscopic factories—"skill" could be more accurately characterized as the skills necessary to match natural materials to industrial production. Workmen were essentially knowledgeable about the natural world, and factories were inevitably connected to it. That reality made industries that processed natural materials more chaotic and organic than rational and mechanical. It was this connection to nature that air-conditioning engineers sought to break through the adoption of new technology and new standards. However, if a close look at the practice of engineering shows an unsurprising hostility toward labor, it also reveals an ambivalence toward management. Greater authority for engineers was achieved by nibbling away at the traditional prerogatives of both management and labor. The engineer's struggle for control over factory production was allied to that of management but not identical to it. As long as each system was the product of engineering design, custom-made for each factory, the division of power over factory production became a three-way split between workers, engineers, and managers. With that larger voice in the direction of the factory, engineers promoted a technology that had embedded in it a quantitative approach to life and one that reordered the factory around those values.

CHAPTER THREE

Defining the Healthy
Indoor Environment, 1904—1929

*T*he air-conditioning industry had an answer for the questions of what air conditioning was (temperature and humidity control), what it should achieve (artificial climate), and who should control it (the technical community). But other groups also recognized these as crucial questions and were not content to let technical experts shape the new technology unchallenged. If air conditioning could produce indoor climate, soon the question arose, what type of climate should it strive to produce? Inevitably, different groups provided different answers to this question. Besides engineers, then, manufacturers, workers, industrial reformers, public health officials, educators, and laboratory researchers all promoted alternate conceptions of the ideal environment.

An informal group of reformers who called themselves "open-air crusaders" challenged the modernist vision behind air conditioning by questioning the new technology at its roots. They attacked mechanical ventilation altogether and sought to establish natural models for the ideal climate. They promoted the open window as the most appropriate technology and community control as the best authority. The open-window approach they promoted was the direct opposite of the ideal of man-made weather. By threatening to capture one of the industry's prime markets, public schools, the advocates of open-window ventilation pushed heating, ventilating, and air-conditioning engineers to further explore the relationship between natural climate and man-made weather.

52

Open-window advocates capitalized upon a change in physiologic theory to push for regulatory reform. The engineering community responded to this challenge in much the same way as they did to the informal challenges to their professional authority within the factory in these same years—by a renewed emphasis upon quantitative standards and experimental science.

FACTORY ENVIRONMENTS

The oldest conflict over atmospheric standards was the one between labor and management, two sides that often disagreed on the ideal factory environment. Atmospheric conditions required for efficient processing did not necessarily promote workers' health and comfort. This conflict was recognized long before the advent of air conditioning. In English textile mills it was common practice to use vapor pots to increase humidity levels by creating large quantities of steam. But high humidities were accompanied by high temperatures, and workers suffered. In response, organized opposition by labor groups prompted the British government to regulate industrial practice. The Cotton Cloth Factories Acts (1889, 1901, 1907, 1911) limited the maximum absolute humidity throughout a range of temperatures, a step that ensured the comfort and safety of the operatives without regard for the effect upon processing. However, the values set were fairly high, the use of vapor pots was not prohibited, and in general the legislation eliminated only the worst practices. Several states in the United States adopted the same conservative legislation.[1]

While regulation prohibited the worst practices of the older technical systems, engineers felt that the new air-conditioning installations would dramatically improve the industrial atmosphere for workers by providing an alternative to basement workshops. Before air conditioning, the relatively cool, moist, and moderate conditions that prevailed in basements made them favorite sites for businesses that needed to escape summer heat. Chocolate dipping was just one of these. If chocolates cooled in a room above 68 degrees, manufacturers feared the fat would rise to the surface of the candy and create an unattractive gray bloom. Some candy makers responded to the problem of discoloration by retreating to subbasement workshops during warm weather, as one critic noted, "in spite of the various disadvantages to the workers and the product that go with the subterranean workroom."[2] Dark, damp, and poorly ventilated, these workshops seemed to encourage tuberculosis. They were just one example of the conflict between workers' health and good processing conditions.

Air conditioning smoothed the transition from basement workshops to aboveground facilities by providing mechanical solutions to environmental control. Bakeries also were commonly located in basements, and public health officials worried about their lack of cleanliness and proclivity to breed disease. In Chicago, for instance, the Department of Health sponsored a municipal ordinance in January 1910 to alleviate these problems. Chicago's efforts to restrict cellar bakeries was aimed at protecting both workers and consumers. The ordinance had a wide impact, as nearly half of Chicago's recorded bakeries (581 out of 1,355) were basement operations. Many of these (371 out of 581) were immediately shut down. Union members supported the new legislation. Members of the Bakery and Confectionery Workers' International Union of America lobbied hard among the members of the Chicago City Council for passage of the ordinance and brawled in the streets with the master bakers' henchmen in support of its enforcement. Union officials wrote: "Let those, who favor our efforts educate the master baker, who so far did not hesitate to sacrifice soul, body and intellect of his employes in the dark underground holes. The bakery workers have a right to enjoy the sun and daylight while they are working, just as much as anybody else desires to enjoy these beautiful gifts of nature to humanity."[3]

The 1910 Chicago ordinance specified that bakeries should be clean, well lighted, and well ventilated. But the last provision was carefully qualified: "Provided, however, that it shall not be necessary to ventilate at such time or in such manner that the process of mixing or rising of dough shall of necessity be interfered with or prevented."[4] In the conflict between comfort and process, even the most progressive municipal authorities put manufacturers' needs first and workers' comfort second. The bakery unionists had lost from the beginning, but blinded by the sunshine they had not seen their new predicament yet. The Chicago ordinance was used as a model for similar legislation in New York City in 1911.

If organized labor was able to achieve limited improvements by harnessing governmental power, air-conditioning systems themselves promised to effect a reconciliation between the needs of production and the well-being of workers. The conflict between process and comfort can be seen as one more instance of management-labor disagreement, but in technical terms the clash was often between the imperatives of humidification and ventilation. When these were treated as separate functions, one was achieved usually at the expense of the other. Opening the window for a cooling breeze could cause humidity levels to fall, with disastrous results for the product. In the bakery, for example, dry air caused the dough to "crust," which interfered with rising and changed its texture, while hot, moist air caused the dough to "slime," which made the bread sour. Indeed, when Carrier Engineering Corporation's head of research, A. E. Stacey, picked up a loaf of bread at his local bakery in New Jersey one summer day, he thought he could detect that characteristic sour taste and immediately tried to sell the bakery an air-conditioning system. Air conditioning that controlled both ventilation and humidification offered one way to reconcile the conflicting demands of these functions, and hence to balance healthy environmental conditions for workers with efficient production. One engineer calculated the impact of the new technology in the bakery: "All in all conditioned air has made life more worthwhile for the baker for with 'ideal weather' every day for each of many purposes he can make better bread, better pies, cakes, biscuits and crackers. He can deliver his product more clean and wholesome . . . and with all this he can provide far better working conditions for his employees."[5] Air conditioning's promise of universal benefits is captured in American Blower Corporation's simple statement that since 1881 the company had manufactured air-handling and -conditioning equipment "for the comfort and health of mankind and the efficiency of production in industry."[6]

The ability of air conditioning to reconcile the requirements of production with the health and comfort of workers was evident in the tobacco industry as well, where manufacturers worried about humidification and workers struggled with dust. Willis Carrier was appalled by the atmospheric conditions when he visited his first tobacco-stemming room. The hand-stemming process created large amounts of dust and minute tobacco particles, and Carrier recalled that "I could see only a few feet in front of me, could not tell whether a person was white or black, and could not see the windows across the room even when the sunshine fell on them."[7] Many of the workers were African Americans, and Carrier recalled that "the operatives worked with handkerchiefs tied over their mouths which they used as home made respirators."[8] The new air-conditioning system, installed primarily to create a high-humidity atmosphere for processing, dramatically improved working conditions as well. At the American Tobacco Company's Richmond, Virginia, plant in 1913, Carrier designed a pan outlet that distributed air like a blanket from the top of the room, at 80 percent relative humidity, with a complete change of air every 3 minutes.[9] "The results were wonderful," he recalled. "When we started up the apparatus you could see the cloud of dust move to the floor, just as if it were a liquid being drained off from the bottom."[10] The air-conditioning company was justifiably proud of the cleaner air that the tobacco workers now enjoyed.

Generally, then, process air conditioning improved the environmental con-

African American tobacco workers. These African American women are working in a tobacco-processing plant where air conditioning has dramatically decreased the amount of dust and particles in the air. Air conditioning often provided both improved atmospheric conditions for workers and better processing conditions for production. (Heating and Ventilating Magazine 15 [June 1918]: 85)

ditions in which operatives worked, successfully finessing the process-comfort conflict. The air washer, or humidifier, which was at the heart of the early air-conditioning systems, was responsible for many of those atmospheric improvements. Consisting of a chamber containing a bank of water sprays, it was superb at adding moisture without causing excessive heat, and in simultaneously washing the dust, fibers, and particles out of the factory atmosphere. Indeed, the evaporative cooling of the sprays could often lower both temperature and humidity levels, especially if it was fed with cool well water. In a few industries, such as rayon, chocolate, photographic film, and gelatin capsule manufacturing, where low humidity was critical, engineers designed, and manufacturers paid for, expensive refrigeration machinery to chill the spray water. Thus, air conditioning provided both cleaner and cooler atmospheric conditions.

Air conditioning fit into the category of technical improvements—like scientific management advocate Frederick W. Taylor's concept of a fair day's work for a fair day's pay—that seemed to benefit everyone. Air-conditioning companies argued that improved worker comfort led to increased efficiency and greater profits. A Buffalo Forge Company catalog pointed out, "It is a recognized fact that atmospheric conditions have marked effect upon the comfort and efficiency of a workman. Thus, the maintenance of proper atmospheric conditions within a plant pays big returns in comfort and contentment of the workmen themselves and in increased and better production."[11] Best of all, such mutual gains were automatic. Manufacturers did not have to assess or pay for the benefits of worker comfort, nor did they have to compromise ideal processing conditions. One observer noted that "fortunately for the people who work in them, most of the industries requiring conditioned air can utilize temperatures and humidities close to those normally desired for human comfort."[12]

Indeed, worker comfort was presented to manufacturers as a happy by-product of process air conditioning. The CEC executives were skeptical that manufacturers would pay hard cash for such difficult-to-measure benefits as increased productivity. Instead, the company suggested that an air-conditioning system would help employers with the perennial problem of labor recruitment. A 1918 CEC advertisement showed the African American women workers of a tobacco factory, with copy that promoted "work in a cool, properly ventilated factory this summer."[13] No workers would ever see this advertisement, but employers were meant to imagine the appeal that comfortable conditions would have for a shifting work force. So, one CEC pamphlet argued, "production is maintained at the maximum and labor, instead of being difficult to secure, actually seeks out the Carrier-equipped mill."[14] Another publication expressed that idea more explicitly when it noted that "in certain industries, such as the textile mill, wherein Manufactured Weather is employed primarily for technical reasons, manufacturers have discovered that the improved, more comfortable and more healthful air conditions have attracted the most desirable employees and practically eliminated labor troubles."[15] While the idea that comfort led to productivity was too hard to sell directly to manufacturers, CEC executives judged that they would have more success with the idea that factory comfort would create a pool of willing workers from which employers could hire the most tractable.

Process air conditioning thus hastened the demise of cellar workshops; cleaned persistent flies, dust, and fumes from factory air; reconciled the conflicting demands of ventilation and temperature control; and mitigated the heat of summer production. Yet despite such notable improvements, it did not ensure workers' true comfort. Industrial heat was one of the problems that air-conditioning systems could temper but not necessarily control. Sweaty bodies, high-speed machinery, and chemical processing combined with hot summer weather often caused the heat of the factory floor to far exceed the outside temperature. Thus, early air-conditioning contracts for textile mills often contained the guarantee that summer temperatures inside the mill would not surpass outside temperatures by more than 5 degrees. In some cases of industrial humidification, CEC recorded that the company allowed the room temperature to fluctuate and instead relied upon large volumes of air to remove heat and maintain high humidities. That is, rather than cool the overheated factory air, the company in those cases simply counted on replacing it. Acknowledging the limits of process air-conditioning systems in providing worker comfort, Willis Carrier told a meeting of industrial physicians that "the system of evaporative cooling so successfully used for high humidities in industry is generally unsatisfactory in air conditioning for human comfort."[16]

That the demands of processing were primary, while the comfort of workers remained secondary, became clear on those occasions when the technology did not automatically effect an equitable solution to both problems. One of the most vivid examples of that conflict for CEC came with the installation of an air-conditioning system in Frankford Arsenal, which used black powder. In 1917 CEC was newly organized as an independent firm, and work for the armament industry was important for its survival. Of its first five contracts, installations in textile factories accounted for a respectable $3,234, while ammunition and fuse-loading plants brought in a whopping $36,530.[17] By 1917 sixteen fuse-loading plants had adopted air conditioning.[18]

As in all plants that worked with black powder, the Frankford Arsenal system was designed to keep humidity levels low and the powder dry. When equipping fuse-loading installations at International Arms and Fuse, British Munitions, and the Rock Island Arsenal, CEC aimed for 45 percent relative humidity and a year-round temperature of 75 degrees Fahrenheit.[19] Yet in this case CEC's client was unwilling to stand the expense of the refrigeration equipment that the company recommended. In a compromise that the company regretted for decades, CEC agreed to install the system without refrigeration in order to keep costs down. The result was a system that had only a tenuous hold on producing favorable processing conditions. Having insisted on a system that the company did not recommend, the client then complained about its performance.[20] To keep humidity low, the system allowed the temperature to rise within the factory, producing an extremely hot working environment.[21] This was not a great deal different from older practice; at Picatinny Arsenal, which had no air-conditioning system, officials either stopped operations dur-

58

ing humid weather or heated the blending rooms to a temperature of around 105 degrees Fahrenheit.[22]

The legacy of this unsatisfactory installation haunted CEC up to World War II. Company engineers described it not only in terms of poor atmospheric control and consequent customer dissatisfaction, but also as an unpleasant work environment. Engineers' complaints about the high heat levels in the plant suggest that their general disgust with their client included an annoyance that the promise of better working conditions for workers—the effortless consequence of most of their installations—had in this instance been betrayed. In common with many engineers in America at this time, air-conditioning engineers handling industrial installations derived part of their work satisfaction from the mitigation of poor atmospheric conditions. They enjoyed being outside the usual labor-management strife, providing a technological fix, as they saw it, to complex social problems.

The ideal of a perfect man-made climate thus underlay not only business advertising but also engineers' professional aspirations. While engineers accepted the limits on the ability of air conditioning to produce true comfort conditions within the factory, they had come to expect that, given the transformative power of the new technology, the conjunction of engineering and business would, at a minimum, produce some improvement. For CEC engineers, this installation was galling because it placed the pursuit of good engineering and business concerns so clearly in conflict and betrayed the inherent promise of the technology.

The potential of air conditioning to reconcile the old conflict between the ideal conditions for processing and those for human comfort seemed to promise an end to labor-management conflict over this issue. That old antagonism had necessitated a resort to governmental authority to balance the interests of these two groups of disproportionate power by regulating factory practice. Engineers believed that such appeals would be unnecessary if air conditioning was adopted in its full perfection. But this optimistic expectation that the need for government regulation would gradually fade away was dramatically destroyed by the eruption of controversy over the ideal atmospheric conditions in American public schools.

SCHOOLROOM VENTILATION

Just as state governments legislated the limits of temperature and humidity in the factory, state and municipal authorities routinely established ventilation requirements for theaters, schools, auditoriums, and all interior spaces where large numbers of people gathered. Regulations commonly called for the circulation of 30 cubic feet of air per minute (cfm) per person. The standards set for these "places of public assembly," in effect, constituted a legal definition of the healthy indoor climate.

The call for a constant change of air was based on the theory that the indoor environment was often unhealthy. Stuffy, unventilated rooms caused such symptoms of distress as headaches, nausea, and dizziness. Eighteenth-century hygienists had believed that the unhealthy nature of indoor air was a result of its changing chemical composition due to respiration; large crowds of people in poorly ventilated rooms caused a toxic buildup of carbon dioxide. Thus, for the next century, carbon dioxide levels were used to measure the healthfulness of the indoor environment.

Such theories about the role of carbon dioxide were soon challenged. In 1842 Maurice LeBlanc questioned the toxicity of carbon dioxide, and soon experimental evidence against the theory began to mount. Eventually, it was overset by the German physiologist Max von Pettenkofer who held the chair of hygiene at the University of Munich from 1865 until 1894. Like his predecessors, however, Pettenkofer still concentrated upon the chemical purity of air as the key to its healthfulness. He advanced the proposition that it was not carbon dioxide but an unidentified poisonous by-product of respiration that was responsible for the discomfort felt in poorly ventilated rooms. This undetected agent was soon referred to as "crowd poison."

Theories of air vitiation had an inevitable impact upon ventilating practice, yet how to accommodate an unknown and undetected poison was not readily apparent. Pettenkofer suggested a continuation of the practice of measuring carbon dioxide levels with the understanding that it represented an index of the organic poison produced by respiration. He published a comparatively simple and accurate method of determining the carbon dioxide level in the air. Later researchers established the average amount of carbon dioxide produced by respiration and the amount of fresh air needed to dilute the carbon dioxide to acceptable levels. The practice of diluting potential toxins, then, came out of physiologic theory. Physiologists recommended 50 cfm per person in order to reduce carbon dioxide levels to 6 parts per 1,000 of air.

Although physiologists set an air supply of 50 cfm per person as a standard for good ventilation, the engineering community was never able to achieve that ideal in economically acceptable terms. Ventilation was always in conflict with heating, because it exhausted to the outdoors air that had been heated at considerable expense and trouble. The experience of the Ruttan Manufacturing Company was probably typical.[23] The company specialized in the ventilation of schools and advertised a generous standard of air circulation. Its experience indicated that although fans and ducts could be increased in size to accommodate the hygienic ideal, typical furnaces were unable to adequately heat such large volumes of air. To install a much larger furnace was too costly. Particularly when remodeling an existing building, adequate ventilation meant inadequate heating. The solution adopted was a reduction in air volume from the ideal. Professor W. Ripley Nichols, faced with the ventilation of a Boston school, admitted that reducing the volume of air circulation might lead to "more or less of the school odor" but "to obviate this entirely would require an amount of fresh air which could not be practically introduced into buildings constructed as the . . . school is."[24]

in the face of this reality, the recommended standard came to reflect engineering practice as much as hygienic ideals. In 1880 John Shaw Billings, the noted public health authority, recommended 30 cfm per person as an acceptable compromise.[25] William J. Baldwin in *The Ventilation of the School Room* (1901) explained that this standard "secures what may be termed a *good* standard of purity, without an excessive cost of maintenance."[26]

Ventilating engineers themselves embraced both the standard and the theory behind it.[27] For while this volume of air represented a moderate standard, it was impossible to achieve a circulation of 30 cfm per person without the installation of mechanical ventilation equipment. With this understanding, the engineering community enthusiastically endorsed the health benefits of fresh air. They set up a dichotomy between the healthfulness of nature and the dangers of vitiated indoor air, with mechanical systems as the necessary intermediary. Machines would be the means for restoring the benefits of nature to the overbuilt and claustrophobic modern indoor life.

The 30-cfm standard gradually became codified into state and municipal ventilating requirements. As early as 1896 New York City schools routinely began to install mechanical heating and ventilation in all new buildings. In 1904 an official committee of the American Society of Heating and Ventilating Engineers (ASH&VE) sought to extend those requirements through an organized campaign to pass statewide legislation, beginning in New York. After several unsuccessful efforts to persuade the state legislature to pass a mandatory ventilation law, ASH&VE built an alliance with the New York state superintendent of public schools and successfully lobbied for a ventilation standard of 30 cfm per person for New York's public schools.[28] The legislation was passed in 1904. New York educators celebrated, because for them the "educational climate" extended quite literally to the air of the classroom; New York engineers were equally delighted because mechanical ventilation in public schools formed a highly lucrative engineering specialty. By 1911 the legislatures of several other states had passed similar legislation.[29]

A chorus of critical voices began to swell within a decade of the passage of compulsory ventilation standards. Criticisms of mechanical ventilation extended beyond the issue of its effectiveness to question its healthfulness. The alliance between public school officials and the engineering community, forged during the first wave of regulation, became increasingly strained by changes in physiologic theory and conflicting professional ambitions.

Increasingly, physiologists challenged the assumption that the chemical composition of air was most important for a healthful environment. In 1911 doubts about the existence of "crowd poison" were raised by researchers in Germany and England. Most influential in the United States were the experiments of British physiologist Dr. Leonard Hill, director of the Department of Applied Physiology of the National Institute of Medical Research. His investigations revealed no dangerous chemical change in air due to respiration.[30] Instead, Hill placed great emphasis on the physical attributes of air rather than its chemical composition; temperature, humidity, and air circulation, he maintained, were better descriptors of the ideal indoor environment than carbon dioxide content. It was "skin effects"— the interaction of the air with a person's skin—that led to people's comfort or discomfort.

The engineering community first learned of the collapse of the chemical theories of air vitiation at the 1911 ASH&VE annual meeting. There Dr. Luther Gulick of the Russell Sage Foundation's Child Hygiene Department dropped the bombshell that the chemical composition of indoor air could no longer be considered a reliable guide to air purity and healthfulness.[31] The engineers found Gulick's comments "electrifying."[32] Suddenly, the medical basis for mandatory ventilation standards vanished, just as ASH&VE was pushing to extend their coverage.

The collapse of the old standard was felt almost immediately in the regulatory process. D. D. Kimball reported to the 1912 ASH&VE meeting that his failure to persuade New York state officials to extend mandatory ventilation ordinances to the factory was "due to the confusion wrought by those advocating different phases of ventilation." The single most debilitating moment came "when the volumetric and CO_2 standards were attacked, and temperature, humidity and elimination of dust were separately advocated by others as of equal or more importance than air volume."[33] The loss of public confidence in the volumetric standard, the rule of 30 cfm per person, left nothing in its place to guide policy or practice.

In response to Gulick's revelation, engineers at both the national and local levels began to form alliances with concerned groups to recast the foundations for mechanical ventilation. ASH&VE voted at the 1911 meeting to form a committee on schoolroom ventilation to cooperate with its opposite number at the American School Hygiene Association (ASHA). The formation of the Chicago Ventilation Commission in 1912 marked a similar effort to forge a new consensus on ventilation standards. The commission, appointed by the mayor, was composed of three ventilation engineers drawn from the ranks of ASH& VE's Chicago section, an official from the Board of Education, the director of laboratories in the Chicago Public Health Department, and the chief commissioner of health.

The collapse of the medical basis for volumetric standards exposed mechanical ventilation to widespread criticism and even to fundamental attack. Critics grumbled about "deference to ventilating systems which often do not ventilate."[34] The *Chicago Evening Post*, for example, sarcastically noted:

> We have sometimes wondered if there was any exercise of the mind more purely speculative than the attempts to work out mechanical systems of ventilation. Given a hall or a public building, the problem is to trace a steady stream of graceful flowing arrows from the outside air through the place, including all the nooks and corners, and out again. The more graceful and enterprising the arrows, the better the system. A really good diagrammatic arrow has imagination, a sense of duty . . . Every architect has his quiver full of these arrows, but very few of his clients ever succeed in getting a glimpse of them . . . And in the case of most public halls and school rooms, the carbon dioxide gets so heavy and inert that it takes complete possession of the place and the fresh air arrows scarcely get a peep within.[35]

In response to the criticism of mechanical systems, engineers complained that school officials were shutting off ventilating systems to save on heating bills. As heating and ventilating engineers focused upon users as the saboteurs of their installations, they began to design systems that would force schools to operate their machinery in a prescribed manner. For instance, some engineers eliminated direct radiation in the classroom and placed heaters in the plenum system so that heat could be supplied only if ventilation systems were kept running.

Despite their agreement to cooperate with ventilation engineers, the ASHA committee complained in 1913 that "even when the [existing ventilation] stan-

dards are fully met, the difference between the beneficial quality of indoor and
outdoor air do not seem to have disappeared."[36] In the absence of a volumetric
standard for ventilation, critics turned increasingly to the natural climate as a
model of proper atmospheric conditions and drew an invidious comparison between mechanical systems and nature.

Such a climate of criticism affected even the Chicago Ventilation Commission, which similarly became embroiled in an examination of the benefits
of mechanical ventilation. The commission struggled with the problem of
whether mechanical ventilation mitigated the atmospheric pollution of modern urban centers, as engineers claimed, or whether it fostered the spread of
unlivable and unhealthy interior spaces. Critics roundly attacked the unnatural and unhealthful aspects of urban life in general and the man-made environment in particular. In a 1913 article, Hill linked his own work in physiologic
theory to these ideas:

> Satisfied with the maintenance of a specious standard of chemical purity, the public
> has acquiesced in the elevation of sky-scrapers and the sinking of cavernous places of
> business. Many have thus become cavedwellers, confined for most of their waking
> and sleeping hours in windless places, artificially lighted, monotonously warmed.
> The sun is cut off by the shadow of tall buildings and the smoke—the sun, the ener-
> gizer of the world, the giver of all things which bring joy to the heart of man, the fit-
> ting object of worship of our forefathers.[37]

For Hill, mechanical ventilation did not relieve the fetid air of indoor living but
was the foundation upon which it was constructed.

In a surprising echo of such criticisms, the Chicago Ventilation Commission
took up the issue of basement workshops. Two years after Chicago's regulation
of cellar bakeries, the commission called upon engineers to refuse to equip underground businesses with mechanical ventilation. It was not because mechanical systems would fail to improve atmospheric conditions there, but rather
because mechanical ventilation gave these spaces the appearance of respect-
ability. Their decision implied that such man-made environments were inher-
ently unhealthy. Owners controlled these systems, and the commission argued
that it was too easy for manufacturers to shut down the machinery to the detriment of workers. The politics of the built environment was rarely acknowledged so explicitly, but it formed an important basis for the opposition to mechanical ventilation. It expressed a vast public uneasiness with modern interior
spaces and an unwillingness to cede control of working, learning, and living
spaces to the engineering community.

In a tidy division of responsibility that bowed to the professional sensibilities of its constituent members, the Chicago Ventilation Commission had begun with the charge that the doctors should decide what constituted perfect ventilation, while the engineers should determine how to accomplish the desired result.[38] But that division of professional responsibilities, which expressed the hopes and expectations of engineers for all of their cross-professional alliances, was repeatedly violated. School authorities, public health officials, and social reformers tacitly refused to hand over to the engineering community such an important part of atmospheric control.

What emerged from the collapse of the old physiologic theory was a renewed interest in the benefits of the natural climate and "fresh air." Particularly critical of mechanical ventilation was a group of radical reformers, the self-styled "open-air crusaders." Composed of school officials, public health professionals, and social reformers, this alliance advocated the maximum exposure of school-age children to the healthful influence of the outdoors. Heavily influenced by their experience treating the problems of tubercular children, they came to the debate over mechanical ventilation with strong ideas on nature, the built environment, and health.

In Chicago in 1909 the open-air crusaders established a summer school for thirty tubercular children, jointly sponsored by the Board of Education and the Chicago Tuberculosis Institute, to provide classes outdoors as a way of improving the children's health while continuing their education. Almost all came from the homes of the working poor where bad health was an obvious result of larger social problems. In the fall of 1909 a grant from the Elizabeth McCormick Memorial Fund established a year-round open-air school for sick children, which consisted of an open-sided tent built on the roof of an existing school. Despite Chicago's cold winters, classes were conducted there all year without any form of heating or mechanical ventilation.

Chicago was also the first city to use this model to establish open-window classrooms for healthy schoolchildren. In 1909 two classrooms left their windows wide open and operated without heat; in 1910, in a concession to public opinion, warm and humidified air kept the temperature above 40 degrees. The supporters of the school reported better health and better conduct on the part of the students in the open-window classrooms than in traditional classrooms. "A cool, humid air is soothing to the nervous system," explained William Watt, the principal of Chicago's Graham School. "We feel better and hence act better in right air."[39] The idea proved widely appealing, and numerous open-air schools were established.[40]

*Student in open-air school, Providence, Rhode Island, January 1912. Many
such schools issued their students special-purpose clothing for the extreme
conditions; this student is wrapped in a padded bag to keep warm. Begun for
tubercular children, open-air schools that operated without heat or mechanical
ventilation were extended to healthy children as well.* (Bain Collection, Print
and Photograph Division, Library of Congress, Washington, D.C.)

Thus began the search for the "right air" for America's public schools that lasted nearly two decades. Watt was one of the leaders of a back-to-nature movement in American education that advocated the elimination of mechanical ventilation from the nation's public schools. In an effort to improve "the environment of childhood and youth," reformers sought a wholesome atmospheric environment for children's early learning and promoted the benefits of natural ventilation. A healthful atmosphere took on added importance given the conflicting theories over the role of nature versus nurture in the development of human character. Most of the reformers believed in the formative importance of nurture over a child's biological heritage. Nurture for them extended beyond the social environment to include the physical environment. They believed that by improving the latter one could promote not only a healthier life but also a wholesome character. They conflated the atmospheric environment with social environment. It was in this context that the physical environment, including ventilation, took on such importance.

The role of the physical environment in child development aroused strong emotions. Hill wrote vehemently that "nurture in unnatural surroundings, not nature's birth-mark, molds the criminal and the wastrel. The environment of childhood and youth is at fault rather than the stock."[41] The emphasis upon the importance of environment, with its corollary that the physical environment was capable of improvement, had a powerful appeal for urban reformers. While some officials blamed the problems of the cities upon immigrants and their supposedly inferior genetic stock, and worked for exclusionary legislation, others turned their energies toward improving urban living conditions. In a rapidly urbanizing country with a long-standing ambivalence toward the moral "healthfulness" of urban life, the anti-urban rhetoric of the fresh-air advocates sounded a familiar and appealing strain.

The open-air crusaders compared mechanical systems and nature explicitly and concluded that nature was better by far. Arguing that standards for large volumes of air circulation were arbitrary because they were based on an outmoded theory of the chemical vitiation of indoor air, these school reformers sought to reduce the legal ventilation requirements to 10 cfm per person, a standard well within the capacity of window ventilation. For them, the solution was simply to open the window and let in some fresh air. But the engineers who had provided heating and ventilating systems for public schools since the 1890s were prepared to fight to maintain standards that required mechanical ventilation.

The stage was set for confrontation with the formation of the New York

State Commission on Ventilation in 1913. In that year Mrs. Elizabeth Milbank Anderson, a patron of the New York Association for Improving the Condition of the Poor, conceived the idea of providing a model ventilating installation in a New York public school. Believing that forced ventilation was the answer to stuffy schoolrooms, "she was prepared to spend substantial sums to equip one or more schools to demonstrate what could be done to insure fresh, pure air, just as pure water, pure milk, and pure food are now, to a degree, guaranteed to congested centers of population like New York City."[42] But Mrs. Anderson was informed that the utility of mechanical ventilation itself was in question.[43] Professor George Whipple of Harvard University was inclined to support mechanical ventilation, but he was willing to admit that "something is wrong with our present methods of school ventilation. Of that there can be no doubt. Expensive installations have failed to provide those conditions of indoor comfort essential to efficient work by teachers and scholars."[44]

Swayed by such opinions, Mrs. Anderson agreed instead to fund an investigatory commission. New York governor William Sulzer cooperated with the Milbank Memorial Fund by appointing the New York State Commission on Ventilation, a quasi-official commission, funded by Mrs. Anderson. Its members included Professor Charles-Edward Amory Winslow, associate professor of biology at the College of the City of New York; Dwight D. Kimball, a ventilating engineer at R. D. Kimball & Co.; Frederic S. Lee, Dalton professor of physiology at the College of Physicians and Surgeons; James Alexander Miller, director of the tuberculosis clinic, Bellevue Hospital; Earle B. Phelps, professor of chemistry at the U.S. Hygienic Laboratory; and Edward Lee Thorndike, professor of educational psychology, Teachers College, Columbia University. In the public-spirited character of the investigation, commission members served without compensation.[45]

The commission conducted experiments to determine the health, efficiency, and comfort of various types of schoolrooms. It investigated four common schoolroom conditions: high temperature, odors, drafts, and low humidity. The members studied sensations of comfort, attendance records, and learning efficiency in an attempt to settle scientifically the question of what constituted a beneficial schoolroom environment. The central issue, however, was whether mechanical or window ventilation produced the best climate for learning. The various independent reports commissioned for the study differed on this question, and commission members must have found it difficult to write the summary report. Indeed, in one experiment designed to measure learning efficiency under various atmospheric conditions, Dr. J. Crosby Chapman and Dr.

William A. McCall concluded that "when an individual is urged to do his best," the atmospheric conditions made no difference in the quantity or quality of the student's work. Their findings emphasized the human environment rather than the atmospheric environment. The open-air crusaders, however, were unwilling to separate the physical environment from the social.

The final report of the commission concluded, much to the indignation of the engineering community, that window ventilation best served the purposes of the classroom. The support for window ventilation was bolstered by opposition to the costs of mechanical ventilation. The commission perceived its charge as the determination not only of the "atmospheric conditions most favorable to human health, comfort, and efficiency," but also of "the most efficient and economical practical methods for securing such conditions." Putting the matter that way settled the issue. Window ventilation was cheaper by far than any mechanical ventilation system. The commission completed its work in 1917, but the final report was not released until 1923. It seemed that the weight of scientific evidence would be on the side of window ventilation.

THE ASH&VE LABORATORY

In those years the findings of the New York State Commission on Ventilation were not public, but neither were they secret. The heating and ventilating community knew in advance what the final report would say, and they were well aware that they had lost a highly visible round of the ventilation fight. In this battle between experts the engineers countered with scientific research of their own. Their experimental data in support of mechanical ventilation largely came from the ASH&VE Research Laboratory, established in 1919 at the U.S. Bureau of Mines in Pittsburgh, Pennsylvania. The laboratory served diverse needs within the profession. One of its most critical contributions was to bolster public confidence in the ventilating engineers' expertise at a time when other professionals were ranged in opposition to them in the regulatory battles.

An authoritative voice for the heating, ventilating, and air-conditioning engineers could scarcely have come at a more opportune time. The idea of establishing a society-sponsored research laboratory was broached by President Harry M. Hart at the 1917 ASH&VE annual meeting. One of the strongest supporters of the proposal was CEC's J. Irvine Lyle, the president-elect of the society, who declared, "I do not think we can ever have too much research." He argued that "there are many problems to be solved through the use of the research laboratory that no experience would exactly duplicate."[46]

Immediate discussion centered on how to fund the research facility and still maintain professional independence. The membership was split on the issue of whether corporate contributions would compromise the work of the laboratory. The engineers turned to the possibility of government sponsorship as one option for financing the laboratory. Members knew that the German government maintained an engineering research institution "for the advancement of the heating and ventilating art." Lyle reminded the society of U.S. government research in agriculture, while another member cited the successful efforts of the American Society of Refrigerating Engineers (ASRE) in securing a $15,000 congressional appropriation to finance research at the Bureau of Standards. It seemed to society members that every other group was more successful at rallying government support than the American heating and ventilating engineers. The balance of concerns between cost and independence was easily tipped, then, when the director of the U.S. Bureau of Mines offered the use of facilities in the newly constructed experiment station in Pittsburgh.

In April 1919 the society chose John R. Allen, dean of engineering at the University of Minnesota, to direct the new laboratory. Lyle served as the chairman of the Research Committee during the laboratory's organization and early development. When Lyle solicited the other members of the committee for their ideas about the direction of the laboratory, the responses reflected the diversity of members' concerns. Frederick Still, associated with a large manufacturing company, proposed a thorough test of the conductivity of various building materials, methods for distributing fresh air, and the most economical water temperature for hot-water heating plants. In contrast, Perry West, a consulting engineer to the New York public schools, was certainly thinking of the ventilation controversy when he argued that "the first and foremost work of this laboratory would be to dispel some of the doubts and misgivings which are in the minds of the public generally regarding the value of modern heating and ventilating and air conditioning."[47] As it took shape, the laboratory attempted to address both technical issues and the larger problem of public confidence.

The laboratory established numerical values for some of the most commonly used constants and improved the instruments used for measurement. Test instruments were often bulky as well as inaccurate. Allen complained that although the anemometer, commonly used to measure air flow through registers, was an extremely valuable instrument when used to impress the uninitiated, he himself preferred to wet the back of his hand and hold it in the air stream when really accurate results were called for.[48] In response to the need for accurate yet practical instruments, F. Paul Anderson supervised the development of a dust determinator that eliminated the use of microscopes and laboratory analytical operations, "most of which required a scientist of high degree for accurate observation."[49] Instead, the new instrument could "be used in the field by any engineer," Anderson claimed, and because it measured relative dust levels, "you might tear up one of your wife's sheets for filters."[50]

The drive for quantitative accuracy was fueled not only by the need for accurate information on which to build effective designs, but also by the desire to supply engineers with the surety of quantitative values in the rugged debate before the public in general and regulatory agencies in particular. One observer noted that as the engineer demanded a larger voice in social decisions,

he must live, henceforth, in a glass house, the eyes of the world upon him. And, perforce, he must clothe himself in circumspection, in reasonable surety of fundamental accuracy. He must not misstep, and jeopardize the hard-won public confidence. Above all, he must set up for himself precise standards of weights, of measures, of procedure, of conduct, of reasoning and conclusion . . . Hence, one of the Engineer's most effective safeguards against loss of public confidence is the development of fixed, rational standards, and the determination of precise finite values, that he may thus express himself with an exactness which ensures understanding.[51]

The laboratory was a good means of establishing just such constants and standards. Clearly, in the debate over ventilation legislation, it enhanced public confidence in the society's claims of representing scientific truth rather than industry advantage.

Under Allen's direction, ventilating engineers began to move beyond their old, strictly reactive relationship to physiologic research. "In the question of the essentials of ventilation, engineers are still awaiting a definite lead from the chemists and physicists," one writer had noted in 1917.[52] In response to the lack of a paradigm to replace the discarded chemical theory of air, the laboratory quickly took up a study of human comfort. Allen followed the lead of Professor John Wilkes Shepard, whose 1916 Comfort Zone Chart defined human comfort in terms of combinations of temperature and humidity. However, Allen's promising beginning at the laboratory was cut short when he succumbed to pneumonia.

Leadership of the laboratory fell to Anderson, who was well aware of the role that quantitative data played in enhancing the scientific character of heating and ventilating engineering. He declared that "the heating and ventilating engineer resents the philosophy that 'the winds will provide.' He insists on controlling his energies in the same intelligent manner as does the hydraulic, telephone, civil, electrical and mechanical engineer. The heating and ventilating engineer is, first of all, a scientist."[53] With such clever juxtapositions, he put science clearly on the side of engineers and not on that of the advocates of natural ventilation.

In addition to strengthening the professional image of heating and ventilating engineers, Anderson brought a certain bonhomie to the operation of the laboratory. He was a well-liked character who was tireless in visiting the local ASH&VE branches to raise money for research. While rejecting the antitechnical rhetoric of Dr. Leonard Hill's published writings, the laboratory under Anderson moved toward a sympathetic testing of his claims that skin effects, such as temperature, humidity, and air movement, constituted the proper basis for determining the ideal climate. At the same time, Anderson could not resist staging an experimental "duel" between two of the society's theorists. E. Vernon Hill, of the Chicago Health Department's Ventilation Division, enthusiastically supported the theory that the wet-bulb temperature of the air was the best determinant of human comfort. Willis Carrier disagreed, noting that the human body seldom acted as a perfect wet bulb. The two opponents were awarded the sobriquets "Wet Bulb Hill" and "Dry Bulb Willis," and under Anderson's supervision, the two met at the laboratory to settle the dispute experimentally. Both Hill and Carrier donned longjohns and stocking caps in an attempt to represent a wet-bulb thermometer. Hill was then doused repeatedly in a cold shower to simulate the wetted wick of a wet bulb thermometer, while Carrier remained dry. Hill quickly abandoned his wet-bulb theory of human comfort. As one investigator recounted the story: "The tests proved that the wet bulb idea literally was 'all wet.' 'Wet Bulb Hill' shivered like a fish, while 'Dry Bulb Willis' stood comfortable and smiling at his side."[54] Done in jest, the experiment underscored the serious problem of a lack of basic knowledge about human comfort and the hope that the society's laboratory would provide some solutions.

Such hopes were well founded. In 1922, just ahead of the circulation of the New York State Commission on Ventilation final report, the laboratory published its first findings. Named the Comfort Chart, it consisted of a graph of the combinations of temperature and humidity at which most people felt comfortable. The Comfort Chart provided an important counterweight to the commission's final report. By embracing temperature, humidity, and air movement as the essence of the ideal environment, the ASH&VE laboratory could define human comfort precisely in quantitative terms. A quantitative description of the ideal environment freed engineers from the always uncomplimentary com-

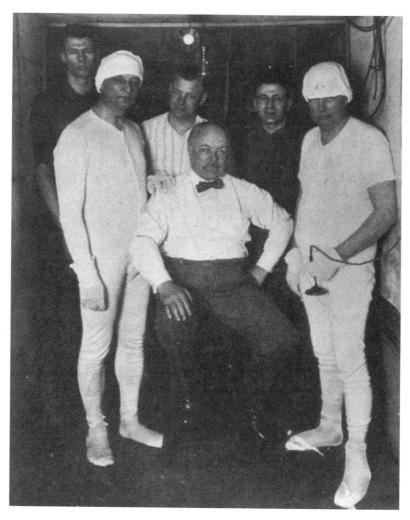

"Research" in comfort at the ASH&VE lab. The director of the ASH&VE Laboratory, F. Paul Anderson, is flanked by Willis Carrier and F. C. Houghten (left), and C. P. Yaglou and E. Vernon Hill (right). Hill and Carrier had different theories about how to measure human comfort, and here they are dressed in clothes that imitate wet-bulb thermometers for a series of playful and bogus comfort experiments. The real comfort experiments of the laboratory helped allay public concerns about the healthfulness of mechanical ventilation and put comfort air conditioning on a more secure footing. Houghten succeeded Anderson as director of the laboratory, while Yaglou did comfort experiments at the Harvard School of Public Health. (*Aerologist* [January 1927]: 15)

parison to nature. In addition, an increasingly scientific content and style of engineering bolstered claims of professional neutrality and authority. The work of the laboratory achieved several of the society's objectives, one of which was to defend its interests against the open-air crusaders.

The determination of standards was not left exclusively to the ASH&VE Laboratory. A significant contribution was also made by the Harvard School of Public Health's industrial hygiene program. Founded in 1918 and reorganized in 1922, the Harvard School of Public Health contained two departments, the Department of Physiology and the Department of Ventilation and Illumination. That division reflected the university's recognition of the importance of both medicine and engineering to the problems of public health. It was the latter department that ran the industrial hygiene program. In 1922 Harvard built a psychrometric chamber (a constant temperature and humidity room) for research into environmental health and comfort.[55] In 1925 the university hired C. P. Yaglou from the ASH&VE laboratory to take charge of the new laboratory. Yaglou provided an important link between the research of the two laboratories.[56] His research at Harvard reflected a continued interest in temperature, humidity, and human health. It was the laboratory at Harvard that subsequently published refined standards for human comfort.

Laboratory experiments enabled ASH&VE to couch the entire air quality debate in quantitative terms. Heating and ventilating engineers attacked the term *fresh air* for its imprecision, and at the 1923 ASH&VE annual meeting the members voted to delete the term from the society's proposed ventilation code. Nothing better illustrates the ideological aspects of the ventilation controversy than this dramatic reversal of language. During the early coalition of public health authorities and engineers, the term *fresh air* had been widely used. Dr. W. A. Evans, Chicago's health commissioner, made it a household slogan in his campaign against respiratory diseases. Evans had drafted Chicago's first compulsory ordinances for ventilation of public buildings and streetcars. His sponsorship of mandatory ventilation and the 30 cfm standard in Chicago was coupled with an aggressive enforcement campaign by the Ventilation Division against theaters and movie houses. Chicago's Ventilation Division provided a model for other progressive cities and an incubator for such advocates of mechanical ventilation as E. Vernon Hill. But Evans's early support for mechanical systems contrasted with his later support of one of the first fresh-air schools in the country.[57] After the work of the New York commission, *fresh air* implied support for the open-air crusaders and an antitechnical spirit.

With a great deal of research having been conducted on each side, the debate over the findings of the New York State Commission on Ventilation was lively. In 1925 Ohio public health officials proposed to reduce the air circulation required in school classrooms from six air changes per hour (approximately 30 cubic feet) to 10 cubic feet, a reduction that brought the level of circulation within the capability of window ventilation. The engineering community vigorously opposed the change. At the public hearings, Chicago-based engineer Samuel R. Lewis testified against the proposal, as did Anderson, the director of the ASH&VE laboratory. On the other side, Dr. E. R. Hayhurst of the Ohio State Department of Health, a hygienist of national reputation, argued for the bill. Testifying at the hearing, Hayhurst cited the findings of the New York State Commission on Ventilation and displayed a copy of the final report, saying, "Gentlemen, this is my authority, this is my bible."[58]

The open-window advocates renewed their efforts in 1926. The New York State Commission on Ventilation decided to extend its investigations with private funding rather than state sponsorship. Its members, with one exception, reconstituted themselves as the New York Commission on Ventilation (the new name reflected the change from state to private sponsorship). The new commission embarked upon a series of tests that charted children's absence from school due to illness in classrooms with various kinds of ventilation. The same year the commission physiologist, C. E. A. Winslow, now at the Yale School of Medicine, campaigned actively for the open window in public schools. He urged readers of the American School Board *Journal* to work for repeal of existing ventilation ordinances, asserting, "Thus, only can the dead hand of outworn hygienic theory be removed and the designers of school buildings freed to provide the most healthful air conditions at a reasonable cost."[59] Winslow succeeded in convincing 15,000 school superintendents in convention at Washington, D.C., to pass a resolution that the temperature of schoolrooms should be maintained at a maximum of 68 degrees and that all "mechanical contraptions" should be dispensed with and open windows with deflectors used for ventilation instead.[60]

To counter the findings of the New York Commission on Ventilation, the engineering community launched its own public relations campaign. Most vocal in defense of mechanical systems was E. Vernon Hill, a former staff member in the Ventilation Department of the Chicago Board of Health. In 1925 Hill founded the journal *Aerologist*. The new journal was quick to point out that in the hot debate over school ventilation physicians overstepped the bounds of their professional expertise. Echoing the Chicago Ventilation Commission's explicit division of authority, Hill complained in an editorial in January 1926 that "the M. D. is contributing valuable and much needed information when he tells us *what* air conditions are desirable in a school room. He makes himself ridiculous when he invades the province of the engineer and tells us *how* to bring these conditions about." Further, physicians seemed to be in a weak position to argue the merits of natural processes. "The physician insists that we be fed treated milk, pasteurized, certified, etc.," Hill pointed out. "He writes volumes on the advantages of filtered and chlorinated water for our domestic use. Why in the name of consistency does he insist upon our using a dirty, polluted, unsanitary air supply, masquerading under the name of 'fresh.' "[61]

Heating and ventilating authorities singled out for special denunciation the dirt and pollution of city air. They declared that there was nothing fresh or natural about the fouled air of modern cities. Only a system of filtration could return urban air to its natural purity. Using mechanical means to achieve or recreate a state of natural purity became a compelling vision for some heating and ventilating engineers.

NATURAL MODELS FOR ENGINEERING DESIGN

If one response of the engineering community to public criticisms of mechanical systems was to put ventilation practice on a quantitative basis through laboratory research, another reaction was to turn more seriously to an examination of the constituent elements of the natural climate. During these years, ventilation engineers explored the natural climate as a model for mechanical systems.[62] Although the *Aerologist* staunchly upheld the superiority of mechanical ventilation over the urban atmosphere, individual engineers, laboratory researchers, and commercial companies intensified their efforts to bring manmade weather more closely in line with the best natural climate. Engineer Frank Hartman explained it this way: "The engineer's conditions are artificial, but because they are artificial, it does not mean that they are not good. There are many synthetic products of the laboratory that are superior to those made in nature's workshop, but when we resort to the artificial to replace nature, we must be sure to include all that nature provided, or our artificially created conditions will fail us."[63]

Some engineers tried to analyze the elements that made up the outdoor climate and to reproduce them inside. Not surprisingly, their natural models were the most pleasant and perfect environs to be found. The beach resort or the mountain spa was most frequently cited as the ideal natural climate. A 1917 text noted that "many experiments have been made and much thought devoted to the problem of ascertaining wherein lies the fundamental difference between

*Ventilation controversy. Here one sees the confusion created by changing physiologic theory in ventilation practice. Fresh-air advocates used the report of the New York Commission on Ventilation to bolster their position that mechanical ventilation should be abandoned in favor of open windows. Air-conditioning advocates were headed in the opposite direction, toward man-made weather. (*Aerologist *[December 1931]: cover)*

the atmosphere of a bracing and invigorating seaside town and a so-called re-
laxing and depressing inland town."[64] Some thought that the wind and wave
action at the beach produced ozone and that this element was the key to bringing the best of the beach indoors.

The most active proponent of adding ozone to ventilation systems was Edwin S. Hallett of St. Louis. Ozone, he asserted, was "that element or condition
of the salt air or mountain valley that is so healing to the invalid and so de-
lightful to all."[65] Hallett, like many of his beleaguered colleagues, was active in
equipping public schools. In 1919 the hygiene department of the St. Louis
Board of Education approached him with its problem. The air in one downtown school was so bad, the department reported, that teachers threatened
to resign on the advice of their physicians. When Hallett visited the school,
he found the student body made up of immigrant children; he noted with
disapproval their garlic-laden diet and their mothers' habit of sewing the children into their clothes for the winter. To counteract the smells of the classroom, he proposed the addition of ozone to the school's ventilation system. His
experiments in the use of ozone to eliminate odors were widely reported and
often copied. While used in this instance for the practical purpose of bringing
the aesthetic qualities of the atmosphere up to acceptable levels, ozone was
sometimes added to the most luxurious installations to achieve an ideal atmosphere.

Other engineers extended the search beyond the composition of air to the
full range of conditions that made the outdoors healthful. Physiologist Leonard Hill's research suggested that the action of natural elements on human
skin was the most important aspect of both indoor and outdoor conditions. In
response, engineers incorporated variable-speed fans in their installations to
reproduce gusts of wind indoors. The Pulsometer Engineering Company sold
a fan that varied from zero to maximum output two or three times per minute
to deliver the air in gusts in imitation of "nature's bracing breeze."[66] Similarly,
when research revealed that glass blocked ultraviolet rays, there was a great
deal of discussion and a few installations of ultraviolet lamps to restore the
healthful effects associated with sunshine.

The growing interest in designing mechanical installations in imitation of
natural climate placed air conditioning at the center of an evolving profession.
As early as 1918 the Chicago Department of Health's Division of Ventilation
voiced the opinion that "the art of ventilation, which occupied a prominent
place in the writings of physiologists and sanitarians a generation ago, is slowly
but surely giving way to a modern science which is usurping the enviable posi-

tion of its predecessor without apology or regret. Even the word ventilation is inappropriate at the present time and is in danger of being replaced by the more expressive, although cumbersome, term, 'Air Conditioning.' "[67]

If engineers who designed process air conditioning were the first to conceptualize their craft as man-made weather, those heating and ventilating engineers who built comfort systems came to the same idea by a different route. Suddenly made vulnerable by changing physiologic theory, these engineers both emphasized the difference between mechanics and nature when they contrasted treated air with urban air, and worked to eliminate that distinction when they expanded the functions of mechanical systems to embrace nature's complexity. This ambivalent relationship between engineering and nature suggests that engineers took seriously the conception of their work as the fabrication of artificial climate.

QUANTIFICATION, EXPERIMENTATION, AND SCIENCE

Of the two responses by the engineering community to the regulatory debate over ventilation, the movement toward more experimental science and quantitative standards was the most far-reaching and long-lasting. On the most immediate level, volumetric standards for ventilation remained on the books generally at the 30-cfm standard. The regulatory battle was largely a contest between competing professionals, as public health authorities and engineers squared off to determine the ideal indoor environment. Since control of an arcane body of knowledge is a defining characteristic of a profession, the scientification of ventilation bolstered engineers' status and authority within the regulatory debates.

Science also transformed relationships within engineering. The establishment of constants for design, the perfection of instruments for measurement, and the generation of standards all turned the "art of ventilation" into a science. While early ventilation and air-conditioning engineering was a rational pursuit, it was not a reproducible science. Laboratory research turned the expertise of individual engineers into a community-held body of knowledge, and that deskilling of engineering shifted power to engineering firms from their technical employees.

Although scientification characterizes all engineering specialties in the modern era, the quantification of comfort that emerged from the ventilation debate was enduringly important for the emerging air-conditioning industry. The medical community's shift in focus to the importance of the physical properties of air played to the strengths of air conditioning. Dr. Leonard Hill's interest in "skin effects" placed the most emphasis on temperature, humidity, and circulation: precisely the aspects of the indoor environment that air conditioning sought to control. Thus the Comfort Chart that plotted the relationship of temperature and humidity was important not only to the ventilation community in general, but also to the air-conditioning specialists in particular.

If the Comfort Chart explicitly defined the ideal indoor environment in terms of complementary levels of temperature and humidity, implicitly it also defined the ideal natural climate. Natural climate could be judged by the number of days that provided temperature and humidity within the comfort zone. The work of the laboratories allowed air-conditioning manufacturers to replace the model natural climate of the seaside town or mountain resort with an ideal climate expressed in quantitative terms. Against such perfection, no natural climate could compete. The shift in standards by which to judge climate, both man-made and natural, was crucial to engineers' success in convincing government regulators to endorse the superiority of mechanical systems over window ventilation. When natural climate was the ideal, mechanical systems sometimes fell short; but when quantitative standards of human comfort became the measure, natural climate was found wanting.

This shift in standards was pivotal for the victory of the technical community engaged in the regulatory battle, but it was important for the commercialization of air conditioning as well. When it was shown that no natural climate could consistently deliver perfect comfort conditions, air conditioning broke free of its geographic limits. When no town could deliver an ideal climate, all towns became potential markets for air conditioning.

CHAPTER FOUR

Motion-Picture Theaters, Human Comfort, and Recirculation, 1911–1930

*M*otion-picture theaters provided a substantial and growing market for the comfort air-conditioning industry from 1911 to 1930. Exhibitors were required, by the same laws that governed public schools, to provide adequate ventilation in their theaters, but their enthusiasm for air conditioning was motivated by more than legal compulsion. Theater chains regularly installed the most modern equipment because air conditioning became an integral part of the luxury and comfort that drew theater audiences. Unlike public school officials who debated the salubrious effects of a return to nature, motion-picture exhibitors embraced art over nature with enthusiasm. More than any other market, the new motion-picture theaters were the place where air-conditioning engineers and their customers happily created the illusion of a perfect artificial climate.

Yet the legacy of the regulatory controversy hit the air-conditioning engineers and their theater clients the hardest. Volumetric standards, which the engineering community fought so hard to preserve in the face of the challenge by open-air crusaders, presented a huge stumbling block to the most economical design in comfort air conditioning. Laws that mandated 30 cfm generally called for all of the air to be outside air, but most comfort air-conditioning installations in theaters relied upon a substantial recirculation of conditioned air to bring operating costs within affordable limits. Air-conditioning engineers engaged in placing comfort air conditioning on a commercial basis for the first time cheerfully violated the sanctity of volumetric standards toward that end.

In contrast, those same engineers strongly supported the findings of professional and academic laboratories on the centrality of humidity control to human comfort. Despite their enthusiasm, however, clients who bought air-conditioning systems and patrons who attended motion-picture screenings were less willing to accept laboratory definitions of human comfort. Once their air-conditioning systems were installed, theater managers ran them to please their patrons, not the engineers.

Thus, the informal settlement of the ventilation controversy was redefined in the marketplace. When making installations in motion-picture theaters, air-conditioning engineers felt free to flout ventilation legislation that mandated 30 cfm of outside air, and theater management ran their systems without regard to experimentally defined standards of comfort. In the marketplace, both law and science received short shrift, and air-conditioning engineers both won and lost in their attempt to gain wider public acceptance of man-made weather.

COMFORT AIR CONDITIONING AND MARKETS

Trying to capture the important interplay between the laboratory and the marketplace, Willis Carrier postulated that "between a fundamental demand for a product and the scientific knowledge of its requirements, the former is the most essential . . . in fact, the all essential factor." But having decided that markets were the most important consideration for the expansion of air conditioning, he felt compelled to add that "on the other hand, an art never achieves its full possibilities, regardless of the market, until the basic principles of its operation are thoroughly investigated and understood."[1] Thus Carrier acknowledged a circle of science and commerce, of research and market. While trying to distinguish which was the most "essential factor," he succeeded only in expressing their indivisibility.

In that sense, then, the research of the ASH&VE laboratory was invisibly but essentially connected to the growing market in comfort air conditioning. The controversy over ventilation ordinances led the engineering community to defend their craft with a quantitative description of human comfort and of the ideal climate. Both were described in terms of the proper combinations of temperature and humidity, and the centrality of humidity control to those new standards clearly pointed toward a future in which the new air-conditioning systems would supplant older mechanical ventilation systems. Only air conditioning could control both elements.

But, as Carrier pointed out, such basic research was useless without a clientele. It was the fast growing popularity of motion-picture theaters that provided the market for comfort air-conditioning systems. The importance of motion-picture theaters in introducing the new technology to the public was clear even to contemporaries. In 1935 one engineer wrote, "It is trite to say that the public became air-conditioning conscious through the avenue of the movie theater."[2] By 1938 an estimated 15,000 of the 16,251 theaters in operation in the United States were equipped with air conditioning.[3] It was in the motion-picture theaters that air-conditioning engineers introduced consumers to comfort air conditioning and tried to educate them to accept this new definition of comfort. Theatergoers responded favorably to the luxury of atmospheric control, but they continued to believe that a comfortable temperature was what made a comfortable room. Air-conditioning engineers repeated the wisdom that it was not the heat but the humidity that made the summer intolerable, yet consumers were slow to accept humidity control as equal in importance to temperature, and slower yet to trade a sensual understanding of comfort for a laboratory-defined ideal.

FROM HEALTHFUL VENTILATION TO REAL COMFORT

The development of comfort air conditioning in motion-picture theaters paralleled the unfolding ventilation controversy described in chapter 3. Inevitably, the evolution of comfort air conditioning was strongly influenced by the public wrangle over the ideal indoor environment. Theaters, like public schools, were required by law to provide 30 cfm per person for the health of their customers. While traditional theaters were built with these concerns in mind, the early nickelodeons were frequently located in old storefronts, and the ventilation was often woefully inadequate to the press of the crowds who filled these makeshift theaters. Storefront motion-picture theaters date back to 1896 but were not widespread until 1906.[4] Nickelodeons took their name from one of the most famous of early motion-picture theaters, the small, ninety-six-seat Nickelodeon Theatre in Pittsburgh, which opened its doors in June 1905 showing *The Great Train Robbery*. It charged a modest 5 cents for admission, as did the many theaters that followed.[5] By 1908 there were 123 motion-picture theaters in Manhattan alone. However, many of those remained small; a typical theater on the Lower East Side, for instance, occupied just 25 × 100 feet.[6]

While theater historians debate whether urban working-class immigrants were essential to the early motion-picture exhibition industry, social historians are adamant that movies were an important entertainment for working-class families. In the growing movement toward the commercialization of leisure, motion pictures supplanted older forms of entertainment based on ethnic and religious affiliations. Motion pictures were accessible and affordable for women, not only the young working women of New York City but even the most house-bound group of all, the married mothers. The low admission price and the family fare of the new motion-picture theaters attracted both men and women, and appealed to many immigrant groups. However, the gathering of the "great unwashed" in one small room inevitably created ventilation problems. Health inspectors in New York City in 1911 reported that in one establishment on Third Avenue, for example, the smell was so vile that an attendant walked up and down the aisles with an atomizer of perfume vainly trying to mask the odor of the theater crowd.[7] While public health authorities believed that all places of public assembly created the potential for disease and discomfort, the nickelodeons were particularly suspect because of their makeshift accommodations and their working-class clientele.

Chicago's public health officials were among the first to marshal municipal authority behind those concerns. When the city's compulsory ventilation ordinance became effective in January 1911, a preliminary survey of motion-picture theaters showed that fewer than 3 percent complied with the new law.[8] Municipal inspectors reported that "these amusement places were being built at a very rapid rate, and it was not uncommon to find small stores, abandoned warehouses, or even livery stables, converted into motion-picture theaters without, in many instances, the slightest attention having been given to the ventilation or sanitary conveniences of the building."[9] The next year brought an inspection of every theater in Chicago, and by 1914 theater inspection absorbed the largest part of the Ventilation Division's energies. The increase in motion-picture theaters in Chicago was astounding, rising from 86 in 1911 to 618 in 1914.[10] A new theater opened roughly every other day in this three-year period.

Their vast popularity and rapid spread resulted in a fierce competition between the early motion-picture houses. As part of that competition to secure customers, exhibitors in New York City began to move out of the small nickelodeons into legitimate theaters that could accommodate a larger audience, and whose location and appointments might appeal more successfully to a middle-class audience. Between 1908 and 1912 New York City's two premier vaudeville theaters, Keith's Alhambra Theatre on Union Square and the Twenty-third Street Theatre, were converted to motion-picture houses. *Variety* noted that "the tendency is toward fewer, bigger, cleaner" nickelodeons. The move into legitimate theaters allowed these more highly capitalized exhibitors to offer a mixed bill of motion pictures and vaudeville, which not only gave them an edge over their competitors but also accommodated both the short length and the scarcity of new films. *Motion Picture News* predicted as early as 1909 that storefront theaters would disappear, replaced "by especially built theatres, seating five hundred to a thousand, most of them giving a mixed bill of vaudeville and motion pictures." Some of the most famous exhibitors built their early successes upon this formula. William Fox became an exhibitor in 1906 and by 1910 owned fourteen theaters that showed mixed programs. Similarly, Marcus Loew began his career in 1904 as a nickelodeon owner and by 1909 operated twelve houses in the New York City area that combined these two kinds of entertainment.[11]

Motion-picture exhibitors routinely installed mechanical ventilation in their new and larger theaters, partly to comply with health regulations and partly to alleviate the uncomfortable conditions of the crowded auditorium. The Poplar Theatre in Philadelphia exemplified that shift. Proprietor A. Stiefel razed his nickelodeon in 1917 and in its place built "one of those modern houses with every up-to-date equipment that has risen phoenix-like on the site of the old nickelodeon." Stiefel seemed anxious to shed the unsanitary reputation of the older class of theater and promised "a disinfecting of the whole once a week." Improved ventilation was part of that new healthful image, and the Poplar purportedly included "perfect ventilation," achieved by means of two large exhaust fans, twenty smaller fans, and twenty doors and windows.[12] The Poplar used the most common type of mechanical ventilation system in theaters, a fan-exhaust system. These systems employed a fan, usually located on the roof, to exhaust warm air through ducts in the ceiling and in the soffit of the balcony. Often built without a separate fresh-air intake, they relied upon open windows or doors for an air supply.

However, many exhibitors hoped their systems would move beyond sanitary conditions to achieve real patron comfort. A 1914 advertisement for the Ziegfeld Picture Playhouse enticed customers with the boast that "sumptuously outfitted and scientifically ventilated, it presents today, and everyday, and every Sunday, the ideal entertainment palace."[13] Mechanical ventilation systems were increasingly expected to provide theater patrons with pleasant as well as healthful conditions. Management at the Robinhood Theatre in Grand Haven, Michigan, claimed that "a system of perfect ventilation makes the theatre one of the coolest in the locality. Exhaust fans in the ceiling and numerous electric wall fans keep the temperature several degrees below that of the street."[14] Such modest claims for comfort on behalf of the fan-exhaust system were vigorously refuted by more disinterested parties. *Architectural Record* referred disapprovingly to the sweep of fresh air through the doorway that accompanied fan-exhaust systems as "the usual draughty opening behind the last row of seats,"[15] while mechanical engineer Edwin Kingsley called the fan-exhaust design "the simplest, the cheapest and the least effective" ventilating system.[16]

In an attempt to alleviate the problems of drafts, some ventilation companies turned the entire distribution system around and provided a supply of fresh air into the theater without an exhaust. The effect of such a system of blowers was to cause "all of the drafts to blow out the doors and windows instead of in."[17] Perhaps one of the most active companies to provide this type of design was the Typhoon Fan Company of New York City, headed by Ernst Glantzberg.[18] The company promised motion-picture exhibitors a lot of comfort for little cost. In a jibe aimed at its competitors, the company claimed that "Typhoons cost no more than buzz and exhaust fans stirring up dust and dirt with noise reminding your patrons of their discomfort."[19] Indeed, a system suitable for a 500-seat theater cost $397, and by 1923 the company had equipped 1,500 theaters.[20] Typhoon fans were installed not only in simple theaters but also in luxury theaters, such as Samuel L. Rothapfel's Rivoli Theatre in New York City. Designed by Thomas Lamb in 1917, the Rivoli was equipped with a perfume system that operated in conjunction with its ventilation.[21]

Most of these mechanical ventilation systems aimed to mitigate the problems caused by people gathering inside on a warm summer day. Densely packed crowds created a room that was hot, humid, and heavily scented in the best of weather; during the summer, the heat and smell of the crowd became unbearable. In response to stifling atmospheric conditions, many theaters were open only during a forty-week season. One Kansas City movie house, for example, was forced to close its doors as early as May 1. Others experienced a dramatic drop in attendance during the summer. On the whole, motion-picture exhibition had a distinct seasonal character. Just as production was seasonal in some industries, so was entertainment.

Some theater patrons found summer amusement in the air-dome, a temporary building open to cooling summer breezes. Proprietor J. E. Barricklow "opened his Airdome where he intends to show pictures during the hot months instead of at his Pike St. theatre," *Motion Picture News* reported. Most exhibitors, however, operated either an air-dome or a legitimate theater, not both. The air-dome was seen as competition rather than as an alternative.[22] The loss of patronage to these summer establishments was associated with a general re-treat by the sweltering public to the outdoors where they could capture cooling breezes. Few people were willing to give up the benefits of the breezy outdoors for the hot stuffy conditions inside.

It is unsurprising, then, that most movie exhibitors equated the ideal theater environment with the best of natural climate, for nature was their stiffest summer competitor. Like public school installations of the same era, theater air conditioning was first measured against the comforts of nature. Theater air-conditioning specialist Otto Armspach argued that "an air-cooled installation is perfect only when the conditions are so uniform and refreshing that patrons are unconscious of the fact that a mechanical source of cooled air supply is present." The ideal he was striving for was "the same feeling of comfort as a similar temperature out of doors on a refreshing June day."[23]

The Typhoon Fan Company was well aware of the comparative delights of the outdoors during the summer and likened its ventilation system to ideal climates. One advertisement promised that "Typhoons give the kind of a breeze that makes seashore and mountain resorts haunted by the comfort seeking public."[24] The claim that its system would equal the most gentle climate was more than enticing rhetoric, for Typhoon executives understood that the lure of the outdoors was the theater's real competitor. Taking aim at the exhibitors' summertime competition, Typhoon assured prospective customers that "the fresh-as-sea air Typhoon Breezes will make your theatre as cool and restful as a sea-shore air-dome—without the mosquitos."[25]

While ventilation companies had long aimed to neutralize the stultifying conditions of a crowded theater, the Typhoon Fan Company strove to market real comfort and not simply to mitigate the discomforts of a hot and crowded theater. In a particularly explicit expression of that ambition, the company assured exhibitors that "folks will come in from the sweltering heat just to 'cool off.' "[26] Typhoon succeeded to some extent with the Old Mill Theatre in Dallas. *Motion Picture News* contended that "it would not be an exaggerated claim that on the most crowded occasion the Old Mill Patron would be a hundred per cent. cooler and more comfortable than he could possibly be at home."[27] Here was a subtle transition from the desire to relieve the discomforts of the crowded room to the ambition to create an interior more comfortable than any of the summertime alternatives.

The industry journal *Motion Picture News* saw mechanical ventilation systems as an important key to summertime competitiveness. The journal noted that "hot weather has a tendency to make people restive and more difficult to satisfy." The experience of the Linwood Theatre in Kansas City was typical. This 1,300-seat theater "drew well enough in the winter, but with summer came a slump in attendance that each year was almost disastrous from a financial standpoint . . . To the amusement parks and airdomes would go the neighborhood patronage in the hot months." The happy solution to the theater's plight was the installation of two large blowers. The journal concluded optimistically that "the history of the Linwood theatre alone bears out the theory that suburban houses can combat summer amusement park competition."[28] The journal lectured its readers:

> There is no real necessity for closing theatres during the summer months. The whole keynote of successful summer business is ventilation. Ventilation can make the theatre a cool and comfortable place, attractive not only by its various forms of amusement but equally, or even more attractive, as an inviting locality to forget the torment of disagreeable, hot weather. Once a theatre has established the reputation of being the coolest place in any given locality, then there is every likelihood that the attendance during the summer may exceed that of other seasons.[29]

AIR-CONDITIONING THE MOTION-PICTURE PALACE

In the growing emphasis upon patron comfort, exhibitors experimented with costly air-conditioning systems instead of the simpler mechanical ventilation systems. These air-conditioning installations were built around an air washer that provided evaportaive cooling by drawing air through several banks of water sprays. The Buffalo Forge Company claimed that "in theatres which are in use during the summer, the air washer provides the means of securing freedom from distressing heat."[30] Consulting engineer Walter Fleisher installed one of the first such systems in New York City's Folies Bergere Theatre in 1911. The Folies Bergere was a dinner theater, and the air washer was purchased specifically to cool diners during the hot summer weather. Fleisher reported that the washer was "not too efficient" but did serve the function of attracting attention to the theater.[31] Apparently, the attempt was not a big enough attraction to keep the dinner theater in business. After opening as the Folies Bergere in April, the theater closed and reopened as the Fulton in October.[32] Most theater installations featured these central-station, spray-type evaporative systems. They were most effective in the dry western states, where low humidity levels produced the greatest amount of evaporative cooling. In 1920 CEC's chief engineer, A. E. Stacey, commented, "In the West, about every moving picture show that considers themselves of class, has an air washer in it."[33] One of the companies most active in installing this design was the United States Air

Conditioning Corporation of Minneapolis. Between 1925 and 1930 its subsidiaries installed 8,000 central-station evaporative-cooling systems, many in theaters.[34]

In the growing movement to create an atmosphere that was better than the outdoor environment, it was not long before engineers added refrigeration to the air washer. This was the only way to gain more control over cooling. Buffalo Forge conceded that "in order to maintain the best cooling effect, refrigerating apparatus for lowering the temperature of the water sprays is sometimes necessary, and may be economically installed and operated."[35] These systems maintained a full range of cooling and dehumidification. It was the appearance of refrigerated air-conditioning systems that finally allowed theaters to move beyond mitigation of discomfort to the creation of an attractive alternative to outdoor amusements. The theater atmosphere was more predictable than the cooling summer breezes of the beach, park, or amusement park. As one Chicago engineer noted, the "escape from the heat of the street offers an inducement for people to step in and obtain respite."[36]

The first documented theater to incorporate refrigeration appears to have been the New Empire Theatre in Montgomery, Alabama, which incorporated a 12-ton refrigeration system with an air washer.[37] Reputedly installed by the American Blower Corporation, the system was in place by October 1917.[38] The addition of refrigeration may have been prompted by Montgomery's hot, humid summers that prevented effective evaporative cooling by an air washer. By later rule-of-thumb standards, which prescribed 70 tons per 1,000 people, this 900-seat theater was about 51 tons short in cooling capacity, yet it reached toward a level of comfort unattainable on the streets.[39]

Thus by 1917 new theaters were being equipped with a variety of equipment: fan-exhaust, blower, evaporative-cooling, and refrigerated air-conditioning systems. Although these types of installations represented a progression of design, in 1917 no single type was dominant. The Poplar, the Rivoli, the Folies Bergere, and the New Empire were all typical installations, though each was different from the others. But 1917 was a watershed year, for increasingly the choice narrowed to refrigerated air-conditioning systems versus all other types.

The 8,000 central-station evaporative coolers sold by the United States Air Conditioning Corporation between 1925 and 1930 suggests that the competitive mix of technologies lasted a long time, but by the 1930s air conditioning with mechanical refrigeration had become the quintessential comfort air-conditioning design. The dominance of that design was strongly connected to the engineers' developing conception of the technology, the consumers' notions of comfort, and the evolution of the technology to meet these expectations affordably.

As was true in factory air conditioning and school ventilation, the designers of comfort air conditioning began to elaborate a conception of air conditioning as man-made weather. First, the health issues that dominated the ventilation laws gave way to a concern for comfort. Then, increasingly, comfort was redefined as an improvement on the outside weather. The theater, an indoor environment, was no longer considered the worst place to spend a summer afternoon but the best place.

By 1917 the idea of creating an indoor space that surpassed nature was linked to the economic competition between motion-picture theater chains. Motion-picture exhibitors promoted comfort as an integral part of a package of luxury, exoticism, and entertainment designed to lure customers. Costly air-conditioning systems soon became a necessary part of their competitive strategy. The way in which competition between exhibition companies fed the growing dominance of refrigerated air conditioning can be seen most clearly in Chicago. In October 1917 Chicago exhibitors Balaban and Katz opened the Central Park Theater. In planning the Central Park, the partners had sought to combine technical excellence with architectural splendor. The company hired the architectural firm of George and C. W. Rapp, because it "delivered what the Balabans wanted in theaters—a grand plan executed with taste."[40] The mythology of the Central Park is that Barney Balaban himself conceived the idea of adapting refrigeration to air cooling for the new theater. He took his idea to the Western Cold Storage Company in Chicago, where he had once worked, to ask the engineer there if it was possible.[41] Thus was born, the story goes, Chicago's first fully air-conditioned theater.

But between the idea and the implementation lay some tricky technical problems. One of the prime technical difficulties in developing comfort air conditioning was the safety hazard posed by the use of ammonia as a refrigerating agent. Ammonia was one of the most commonly used refrigerants, but engineers were well aware that it was a volatile chemical that presented the danger of explosion or toxic leaks. The toxicity of ammonia was judged an unacceptable risk around large groups of people; in other words, the risk was greatest in just those applications which could most benefit. These safety problems could be circumvented by remotely cooling a brine solution and circulating it through a cooling coil in the air path of the fan, but one engineer calculated that the expense involved in dropping the indoor temperature from 90 degrees to 70 degrees purely with cooling coils without an air washer was "impracticable commercially."[42]

These problems were solved, not by Barney Balaban or anyone at Western Cold Storage but by Chicago engineer Frederick Wittenmeier. Wittenmeier was chief engineer for the Kroeschell Brothers Ice Machine Company from 1897 until 1917 and then headed the Wittenmeier Machinery Company from 1917 until his death in 1928.[43] He found a safe and economical solution to the problems of comfort air conditioning through the use of carbon dioxide as a refrigerant. Carbonic refrigeration machines were widely used in Europe but little known in the United States before 1900. Wittenmeier made several contributions to the technical refinement of early CO_2 machines in the United States, but his primary contribution was the commercial development of carbonic refrigeration in the United States.[44]

Under Wittenmeier's leadership, Kroeschell Ice Machine Company installed several early carbon dioxide fan-coil cooling systems, most notably in the Pompeiian Room and banquet hall of the Congress Hotel in Chicago in 1907, and in Frank Lloyd Wright's Larkin Building in Buffalo in 1909.[45] Then, sometime before October 1910, Wittenmeier took advantage of the safety and versatility of carbon dioxide to design an air conditioner that placed direct-expansion cooling coils within the spray chamber of the air washer itself. He filed patent applications for his designs in October 1910 and May 1911, assigning the rights to his employer Kroeschell.[46] Wittenmeier's incorporation of refrigeration coils into an air washer was explicitly aimed at the control of cleanliness, temperature, and humidity—in other words, at the development of an air-conditioning system.[47]

It is equally clear that he expected theaters to be an important market for his new system. A Kroeschell catalog explained that "modern people demand comfort and for this reason air cooling of public spaces has developed a tremendous field for refrigerating machinery using carbonic gas as a refrigerant. The moving picture theatre has done more to develop this field than any other type of institution."[48] The large motion-picture palaces with 2,000 to 4,000 seats often sold the same seats for three or four showings a day; Wittenmeier noted that "it makes a tremendous difference in the receipts if the house is filled 80% on average or only 40% or less."[49] With so much capital invested in their theaters, exhibitors could not afford a drop in attendance during summer. Air conditioning would draw in crowds and so defeat the traditional summer downturn. The air-conditioning system in Balaban and Katz's Central Park Theater incorporated many of the features of Wittenmeier's 1910–11 patents and was designed to produce an air temperature of 78 degrees inside the theater when the outside temperature was 96 degrees.[50]

The Central Park Theater opened Saturday evening, 27 October 1917, and was acclaimed for both its beauty and its technical innovations. *Moving Picture World* reported that the theater was "without a doubt the most palatial structure devoted to moving pictures in Chicago, if not in this country," while *Motion Picture News* declared that "the Central Park is a masterpiece." The *Motion Picture News* reported further that "never at any time in the motion picture history of Chicago did a house get the ovations from an opening night audience that was accorded this synthesis of all progressiveness in theatre construction." The two sets of brothers, Barney and Abe Balaban, and Sam and Maurice Katz, "who scoured the earth to bring together the equipment and devices that are found under the roof of this playhouse, were swept off their feet by the congratulatory wave that broke over them." Proclaimed "the acme of Chicago's achievement in theatre construction," the Central Park Theatre brought out motion-picture producers, exchangemen, exhibitors, and theater-supply representatives from all parts of the country who attended on opening night to get ideas on house construction.[51] It was that combination of local and national attention that made the Central Park air-conditioning system so influential. It provided a model for other Balaban and Katz theaters and a point of comparison for Chicago competitors. That pattern was repeated all over the United States. The trend toward multitheater holdings in the motion-picture exhibition industry accelerated the adoption of air conditioning, for a single executive decision could diffuse the technology throughout a national chain. The appearance of air conditioning in a theater belonging to a national chain could spark its rapid adoption at the local level, fueled by the intense rivalry of exhibitors.

The air-conditioning system in the Central Park Theatre certainly gave Balaban and Katz an edge over the firm's Chicago competitors. Barney Balaban recalled that "up until then no one ever thought of going to the theater in the Summer time. But we proved that we could do business fifty-two weeks a year."[52] The air-conditioning system proved so successful that Balaban and Katz made it a standard item in their chain of Chicago theaters, equipping the Riviera (1919), the Tivoli (1921), and the Chicago (1921) with similar systems.[53] The Wittenmeier Machine Company proudly advertised, "Cooling and Dehumidifying the air during the summer makes a theatre equipped with Wittenmeier System a profit producer."[54] Chicago engineer Samuel C. Bloom expressed more fully the economic benefits of air conditioning: "Good pictures alone cannot be relied upon to produce satisfactory revenues. The cooling of such places in hot weather enables the theatres, particularly motion picture houses, to show to full houses continuously."[55]

The Central Park installation inspired not only its owners but also competitors to copy its air-conditioning system. When the Orpheum chain opened the 2,260-seat State-Lake Theatre in Chicago in 1919, it hired the old Kroeschell firm, now the Brunswick-Kroeschell Company, to install a carbon dioxide air-conditioning system with 250 tons of cooling capacity.[56] The State-Lake provided the Orpheum Company with a model for its chain of theaters. "The great State-Lake Theatre in Chicago . . . was a new type and since its phenomenal success, four others . . . have been built along the same general lines," one observer reported.[57] The Orpheum theaters in Los Angeles and Kansas City, equipped with carbonic systems, kept the focus of local competition on atmospheric comfort.[58] In 1928 one Los Angeles engineer reported that "we have two houses—the Orpheum and the Metropolitan—that have adequate refrigerating systems, and as an index of the prevailing trend, three more are in course of construction."[59] New York engineer Stephen Bennis observed the same trend, noting that "especially in New York City competing theatres are putting in air conditioning apparatus for profit."[60] Most air-conditioning systems were not a direct response to climate. Instead, they more nearly reflected the fierce economic competition in a rapidly growing industry that lured in potential customers by offering irresistible luxuries and hoped to do so year-round.

The Central Park represented a new strategy in motion-picture exhibition that emphasized the appeal of the theater itself. One theater professional observed that "it may well be said that the modern theatre is the playhouse of the masses."[61] Others, however, saw the appeal of motion pictures as universal, and the motion-picture theater as a bridge between classes. Theater architect George Rapp called the new picture palace "a shrine to democracy where the wealthy rub elbows with the poor."[62] Indeed, Balaban and Katz promoted their theaters as the most democratic of institutions. One advertisement explained, "If we attempted to establish financial class distinctions, or to divide our audiences by means of reserved sections which seem to be more desirable and exclusive, we believe we would be destroying the plan which made these great theatres possible . . . These theatres are for 'all the people all the time.' "[63] Despite this classless and seasonless rhetoric, it was not the leveling of social barriers that appealed to the general public but instead the mystique of the rich and famous. One observer noted that "theatre owners, almost universally, are of the

Balaban and Katz theaters. In advertisements like this one for the Balaban and Katz Central Park Theatre in Chicago, motion-picture theaters competed for audiences by emphasizing good entertainment, exotic architecture, and a comfortable atmosphere. Air conditioning ended the seasonal character of movie exhibition and established theaters as the most comfortable place to be during hot, humid weather. (*Chicago Daily Tribune,* 23 June 1919)

opinion that the average theatre-goer comes to the theatre to get the thrill of rubbing shoulders with the elite and basking in luxuries that their homes cannot afford."[64]

Although exhibitors could not necessarily deliver the city's social elite, they could and did build architecturally extravagant theaters to create an appropriate setting for even the very wealthy. One theater architect explained that "these large over-ornamented theatres, commonly found in all big cities, are intended to create a sumptuous feeling, and to intrigue the beholder with their glittering gold cornices, huge imitation marble columns, and excess of ornamental plaster, used in their lobbies, foyers, stairways and auditorium. Thus, the theatre industry has arrived at a combination of an exhibition of palatial architecture and motion pictures."[65] *Motion Picture News* complained that theaters were now more important than the movies themselves, adding grumpily that "a bid for patronage has become a competition, not between the merits of film offerings as much as between lobbies."[66] Marcus Loew, who presided over a nationwide chain of theaters, captured the growing importance of the theater's architecture and appointments more simply when he declared, "We sell tickets to theaters, not movies."[67]

Increasingly, atmospheric conditions formed an important part of a theater's attractions. Comfort became an indivisible part of the exhibitors' claims to luxury. Loew's construction engineer believed that "the average theatre audience are never ready to pay admission unless you offer them the three and only important privileges they are interested in: A GOOD SHOW, A GOOD SEAT AND COMFORTABLE AIR CONDITIONS."[68]

DEFINING HUMAN COMFORT IN THE MARKETPLACE

The Central Park Theatre and its fellow, the Riviera, focused discussion within the engineering community on how to define comfort. If Willis Carrier was right in saying that the successful exploitation of markets depended upon a knowledge of fundamental principles, then air-conditioning engineers were certainly in trouble. For at the time these theaters were designed, few engineers could say with confidence that they understood what constituted a comfortable room. On 4 July 1919 the temperature outside the Riviera Theatre stood at 94 degrees, while the inside conditions hovered around 74–78 degrees and 70 percent relative humidity.[69] Was that comfortable?

Comfort was such a subjective sensation that engineers fell back on anecdotal evidence to define its limits. The extremes were not hard to identify. Sausage manufacturing, for instance, required colder temperatures and damper conditions than workers liked; one engineer recorded that complaints began when the temperature dropped below 55 degrees and increased in number as it approached 50 degrees.[70] But the more moderate conditions were less clearly defined.

Not surprisingly, Wittenmeier's understanding of comfortable conditions was based on the pathbreaking 1907 installation in the Congress Hotel. The management had asked for a system to deliver 70-degree air in the dining room, but within a week diners were complaining of the cold, and the thermostat was changed instead to 73–74 degrees. Based on this experience, Wittenmeier felt that inside temperatures should be set in relation to the outside temperature. As the outside temperature climbed, so should the thermostat; the most comfortable temperature, Wittenmeier claimed, was a 15-degree difference with the outdoors.[71]

The strongest criticism of Wittenmeier's theater air-conditioning systems concerned not the temperature but the high relative humidity. "Temperature was all right," CEC engineer L. L. Lewis explained, "but humidity was all wrong."[72] The theater felt damp, and moisture condensed on the seats, walls, and draperies. Others remembered that "the system was designed to give low temperatures irrespective of relative humidities, and little thought was apparently given to proper relation between both of these factors as required for comfort."[73] His patented air conditioner provided for humidification but lacked an effective method for adjusting the humidity level. Air left the air conditioner nearly 100 percent saturated, and the body heat of the audience raised it about 8 degrees. In the Riviera, that produced a relative humidity of approximately 70 percent. Wittenmeier routinely furnished a temperature of 76–78 degrees and a relative humidity of 75 percent. "I assure you that you will feel comfortable in such a house," he maintained.

Wittenmeier was less concerned with humidity levels than was Carrier, whose success was based upon precise humidity control for industrial plants. Yet not all of Carrier's industrial installations furnished useful insights into the proper atmospheric conditions for comfort. For instance, Carrier once watched the behavior of workers in a textile plant equipped with process air conditioning but could discern no rhyme or reason to their decisions to lunch inside the conditioned mill on some days and outside on others. However, in fuse-loading plants during World War I he discovered that when otherwise comfortable temperatures were combined with high relative humidity, factory workers experienced problems. Sweat did not evaporate off their hands, and explosive powder clung to the damp impressions that workers left on the dies.

That residual powder would flash when put into the press, and such incidents occasionally caused a serious explosion or fire.[74] This experience caused Carrier to focus upon the "sweat point" (the point at which sweat failed to evaporate quickly from the body) as one of the keys to human comfort.

Carrier brought to the comfort problem years of experience in industrial air conditioning, where humidity often mattered more than temperature and where precision in humidity management made possible the economic benefits of process control. A concern with precision was part of his experience and an essential part of his engineering expertise. In contrast, Wittenmeier sought to design a system that achieved low temperatures at an affordable price. He understood that lower humidities could be achieved by first lowering the temperature to 50 degrees and then reheating the air before it entered the auditorium, but alternately cooling and heating the air during the summer was wasteful and expensive. In his view, such precise humidity control was "not considered necessary for public buildings and theatres."[75] For Wittenmeier there was a large gap between industrial practice and the new comfort air conditioning.

ECONOMICAL HUMIDITY CONTROL AND RECIRCULATION

Complaints that the Wittenmeier systems were cold and clammy prompted Balaban and Katz in 1919 to turn to Carrier Engineering Corporation for a bid when they planned their next theater.[76] Now it was CEC's turn to devise a way to bring down the humidity levels and still produce an economical system. It was not an easy task, and members of the New York office voiced their fear that their design would be "as thoroughly unsuccessful as [the buyer's] previous installation."[77] Part of the problem was Chicago's ventilation requirement of 25 cfm per person. When engineers introduced that amount of cool air into the theater, the typical body heat of each patron would raise the temperature of the air no more than 7 or 8 degrees. These levels created a low temperature and a high relative humidity in the theater that caused a feeling of discomfort. At a time when the heating and ventilation industry was busy forming alliances with other professional groups to save the existing volumetric ventilation legislation as a defense of mechanical systems, those standards were blocking the evolution of mechanical ventilation into comfort air conditioning. The air-conditioning minority struggled to find a design that would economically adapt industrial practice to comfort air conditioning, and still stay within the limits of ventilation legislation.

After consulting with both J. Irvine Lyle and Carrier, the New York office presented three solutions. The first option was to reheat the chilled air as it came out of the air washer in order to bring the relative humidity down to comfortable levels. This was a common feature of industrial systems where precision was all-important, but it was too expensive for most customers of comfort air conditioning and was the easiest option to discard. The second possibility was to reduce the air circulation to 10–15 cfm per person so that the body heat from the audience would be sufficient to raise the temperature and thus lower the relative humidity to an acceptable level. This, however, would require a special dispensation from the Chicago Health Department, which legislated 25 cfm per person. The third alternative was to continue to provide 25 cfm per person, but to design a system in which half the air supply was made up of air recirculated from the auditorium of the theater.[78] Technically, this did not meet Chicago health standards either, for the Ventilation Division specifically required 25 cfm of *outside* air. Such recirculation provided no more than 12 cfm of fresh conditioned air, comparable in volume to the second option. Wittenmeier had used this third option in the design of the Riviera Theatre. That system returned 50 percent of the cooled and filtered air in the theater to the inlet to be mixed with the outside air supply before being pulled through the air conditioner a second time.

However, Carrier's commitment to humidity control eliminated this type of recirculation as a viable option for CEC. Instead, the firm proposed a little-used type of recirculation, which it eventually called a "by-pass system." The bypass system appears in the company's plans for the Balaban and Katz theater. Unlike Wittenmeier's design, this system would not send the return air through the air washer and recool it, but would mix it with the cold air that came out of the air washer in order to raise its temperature and lower the relative humidity before the conditioned air entered the theater. In effect, the company eliminated the heater of industrial practice and instead raised the temperature of the conditioned air with warmer air from the theater. This technique of recycling the air of the theater and mixing it with the treated air before returning it to the auditorium received its name because the recirculated air bypassed the air washer. As early as 1919 CEC identified this as a possible method of achieving economical comfort air conditioning. The bypass technique preserved humidity control as an important part of comfort air conditioning and thus linked the company's existing expertise to a growing new field.

Despite this early recognition of its potential, CEC's first bypass system was not installed until 1921, in Sid Grauman's Metropolitan Theatre in Los Angeles. L. L. Lewis sketched out his ideas for the system in a company memo of 9

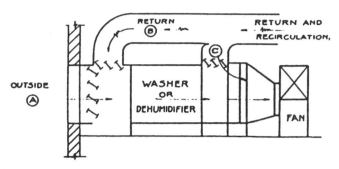

*Bypass. As depicted here, in bypass circulation, air is returned from
the air-conditioned room and bypassed around the air washer at* C.
*This warmer, recirculated air raises the temperature of the treated
air to bring it up to a proper level of humidity and comfort before it
is reintroduced into the room. Bypass circulation proved to be the
most economical way to achieve humidity control in comfort air
conditioning, and consequently the patent holders captured an
estimated 90 percent of the comfort air-conditioning market.*
(*Refrigeration Engineering* 15 [May 1928]: 122)

September 1921 to E. P. Heckel, who went to Los Angeles to estimate the job.[79]
Then, on 22 December 1924, Lewis filed a patent application on his ideas.[80] Essentially, the system supplied the full 30 cfm per person mandated by law, yet only 25 percent of that air was outside air that had been washed and cooled, while 75 percent was air recirculated from the theater auditorium that bypassed the air washer. The reduction in the volume of air sent through the air washer meant that the refrigeration machinery, an expensive part of the installation, could be greatly reduced in size.

The Metropolitan Theatre epitomized the combination of innovative mechanical systems and extravagant architecture. It was Grauman's third large Los Angeles theater, built after the Chinese Theatre and the Million Dollar Theatre (1919). Designed by architect William Lee Woollett, it has been described as an "exotic jumble of Middle Eastern ornament and geometric patterning."[81] Three years in the planning, it was part of Grauman's ambitious plans for a theater–office tower complex on Pershing Square that were never entirely realized, despite the formation of a partnership with Paramount Publix to give him greater financial resources.[82] Nevertheless, the house that opened in January 1923 was spectacular for its size (3,500 seats), its lavish decoration, and not least, its air-conditioning system.

The Metropolitan was expensive, reputedly costing $3 million, and one newspaper reported that the air-conditioning system accounted for $115,000 of that total.[83] Construction costs were only part of the expense of air-conditioning the theater. Operating costs for the installation ran $500 per month during the winter and rose to $2,200 in the summer, even with bypass recirculation.[84] To some such an expenditure was an essential part of a new theater, for as one engineer noted, "builders of new theatres would no more leave out cooling than they would neglect heating, for the public demands it and experience has demonstrated that it pays."[85] But for Grauman in 1920, when planning for the theater began, including air conditioning was an innovative step. In 1923 he told a Los Angeles publicist, "The public doesn't *demand* anything . . . It is only after a thing is created that the public demands it. If it were possible to consult the public and discover what it wants, the showman's job would be easy."[86] Indeed, Grauman's Million Dollar, Egyptian, and Rialto theaters were all equipped with evaporative cooling systems without refrigeration.[87] The operating engineers at the Million Dollar Theatre created extra cooling during hot weather by carrying cakes of ice on their backs to deposit in the water tank of the air washer. Despite the high costs of purchase and operation, Lewis calculated that in any theater an increase of slightly less than 2 percent in attendance would pay the cost of an installation.[88]

If CEC struggled to make air conditioning affordable, the company also attempted to make it comfortable. By controlling the relative humidity, the company believed it would be possible to allow the temperature to rise and still achieve comfortable conditions. Indeed, CEC let temperatures rise as high as 78–80 degrees in its installations but kept the relative humidity between 45 percent and 55 percent. This combination appears to reflect Carrier's industrial experience with the "sweat point." The firm claimed that under such conditions "a person will not feel a shock entering from the outside and yet the perspiration will be gradually removed from the body and from the clothing."[89]

To achieve greater comfort, Lewis also designed a new distribution system for the Metropolitan Theatre that was specially adapted to air conditioning. Many theaters traditionally used an "upward" method of air distribution that proved poorly suited to the new technology. In such a system, air was introduced into the auditorium through "mushroom" ventilators (so named because of their shape) located underneath the seats. Theoretically, the air rose to the breathing zone where it was gradually warmed by body heat. This vitiated air rose upward and was exhausted through ducts in the ceiling by an exhaust fan. It was then replaced by fresh air rising from the mushroom ventilators. "The upward method," one fan manufacturer wrote enthusiastically, "is to be desired wherever the architectural design makes it permissible . . . the air flows upward in accordance with the natural air currents induced by the heat of the body and the breath."[90] As late as 1922 CEC installed just such a system in the $2 million Granada Theatre in San Francisco.[91]

For the introduction of cold air, however, this method proved disastrous, for the cold air never rose into the breathing zone, but settled on the floor. Air-conditioning engineers noted with dismay that moviegoers carried newspapers into the theater to wrap around their feet during the show.[92] To counter this problem, Lewis installed a distribution system that introduced the conditioned air from the ceiling, which wags promptly dubbed the "upside down" system. Officially, the system was called a down-draft method of distribution.

The Metropolitan down-draft system was not a complete success. Lewis joked that the new down-draft distribution system exposed the engineer to criticism from "all of the types from red-heads to bald-heads."[93] And apparently he received plenty of criticism over this system. He admitted later that "Grauman's proved the generally accepted point that the first of revolutionary jobs is not always perfect."[94] For *Boxoffice* magazine, Lewis explained the problems of the Metropolitan discreetly: "Factory experience found itself in the anticipated conflict with architectural and decorative features, and handicapped by the obstacles which they imposed," he wrote.[95] The problem appears to have been a reluctance by the architect to use distributors, trumpet-shaped outlets that diffused the cold air gently over a wide area. Distributors proved effective in factory practice, but architects found their appearance objectionable in these elaborate movie palaces. In a memo to company engineers Lewis described the conflict more bluntly: "I doubt if you will work on a job with any architect or engineer who will not, sooner or later, suggest that we blow directly downward through a grille work. We tried this on the Metropolitan, the Palace, the Rivoli, the Texas, the Miami, and maybe one or two other theatres. It has failed completely on all those theatres which we name."[96] Eventually, engineers learned how to reconcile aesthetics and technical requirements. The pan outlet, a panel hung below the discharge over which the cold air diffused, was one solution. A rival engineer acknowledged in 1926 that "at the present moment the system which is predominating is the downward system of ventilation."[97]

By 1922 CEC had adopted two important innovations for the successful adoption of comfort air conditioning: down-draft distribution and bypass circulation. Carrier recalled that "the successful application of the downward ventilation with independent temperature and humidity control, without the necessity of reheating, made it practicable for the first time in 1922 to air-condition theatres, both acceptably and economically."[98]

Bypass circulation was essential to any economical comfort air-conditioning system. Carrier Engineering Corp. owned Lewis's bypass patent, while rival Walter Fleisher held a patent on a similar design.[99] Despite this legal division of rights, Carrier acknowledged that his firm installed approximately 300 air-conditioning systems that infringed upon Fleisher's patent. In the midst of the resulting litigation over the infringement of bypass rights, Carrier and Fleisher agreed to form a patent pool. In 1927 they formed the Auditorium Conditioning Corporation with the Lewis and Fleisher inventions as a core, and eventually acquired thirty-one more related patents.

The centrality of the bypass patents excited a great deal of resistance to the new corporation. York Ice Machinery Corporation filed the first test case and lost. In 1929 the triumphant Auditorium Conditioning Corporation defined its victory for potential clients:

> It was shown, and agreed by the Court, that the "recirculation" of large volumes of air, by-passed beyond the Conditioning Machine, as in these methods, for the purpose of temperature and humidity control, constituted a new and important improvement in Air Conditioning. Hence, regardless of what may appear to be "different" methods of "recirculation" any form of "recirculation" employing effectively *large* volumes of air recirculated without passage through the Conditioning Machine, may be deemed an infringement of the Auditorium Conditioning Corporation's patents.[100]

The centrality of its system is evident in the fact that by 1946 Auditorium Conditioning Corporation had licensed an estimated 90 percent of the comfort air-conditioning installations in the United States.

HUMIDITY AND HUMAN COMFORT

The work on human comfort begun in 1918 by the newly formed ASH&VE laboratory seemed just what the emerging comfort air-conditioning industry needed to design truly comfortable installations. The changing physiological paradigm, from an emphasis on chemical vitiation of air to a focus on physical properties, was crucial in shifting the goal of ventilation from health to comfort. That comfort became synonymous with health was fortuitous in uniting the engineering profession.

When the first results of the ASH&VE research laboratory were published in 1923, they established the idea of "effective temperature." Varying combinations of temperature, humidity, and air movement produced sensations of equal comfort, or equal effective temperatures. It was, indeed, the experimental proof of the important influence of humidity and air movement on personal comfort. Although the data had been collected largely to influence the regulatory battle over school ventilation, air-conditioning engineers used the information to settle the conflict over comfort cooling. One specialist characterized it as "the greatest contribution to the art of air conditioning in its broadest sense which has appeared in a generation and is worthy of the careful study of every engineer concerned with the effect of air conditions upon human comfort."[101] By 1925 the values of 78 degrees and 50 percent relative humidity were widely used as the basis for comfort.[102]

The effect of the ASH&VE laboratory study was to help transform a personal, sensual experience into an objective, quantifiable standard. For example, city authorities in Minneapolis required the ventilating contractor for the municipal auditorium to guarantee that his installation would produce comfortable atmospheric conditions based upon ASH&VE's Thermometric Chart.[103] Such standards inevitably had an important economic impact. The experimental establishment of the centrality of humidity to human comfort favored firms whose technical expertise, equipment, and patents already incorporated humidity control. Clearly, in the scientific debate, Wittenmeier's high-humidity air-conditioning system lost out to CEC's bypass system. Mikael Hard has called such use of science to adjudicate technical claims by competing commercial firms "the scientification of the marketplace."[104]

The strength of CEC in the emerging comfort air-conditioning field was based on several factors. First, theaters, like factories, required custom-designed systems, and the company's engineering expertise placed it in the forefront of the small group of engineering firms that dominated the industry. Second, that combination of engineering expertise and economic power allowed CEC to promote with some success a definition of the new technology that placed humidity control at its center. This position was bolstered by the wave of comfort research that followed the ventilation controversy. Third, in any installation that required humidity control, CEC held an enviable patent position. The bypass system was important because it was economical, but also because it produced atmospheric conditions squarely within the limits of the comfort zone. The bypass salvaged humidity control as an essential part of the new technology just when engineering practice threatened to abandon it as too expensive for the new markets. Through the Auditorium Conditioning Corporation, CEC seemingly stood astride the future of comfort air conditioning.

However, the scientification of the marketplace was possible only if the public accepted the authority of the engineering experts. Evidence suggests that both theater managers and patrons resisted the formulation of comfort dominated by the importance of humidity control. Theatergoers seemed generally uncomfortable with Wittenmeier's high-humidity–low-temperature systems, but the engineering literature suggests little enthusiasm among theater managers for an alternative low-humidity–high-temperature system. Backed by laboratory findings, CEC engineers were convinced that 80-degree rooms would be comfortable if the humidity were kept to approximately 50 percent, but few customers agreed. Wittenmeier, for his part, had given up on ideal engineering standards of comfort. Despite his own conviction that theaters should maintain no more than a 15-degree differential with the outside temperature, the Riviera Theatre was routinely kept 20 degrees colder than the outside. Most exhibitors wanted to make the most of their air-conditioning systems, and a 20-degree differential sounded better than one of 15 degrees. After maintaining the 20-degree differential in one theater all afternoon, Wittenmeier reported that "the owners of the house thought it was fine; they insisted on having it. They think that is about right for moving picture houses, and we have to be guided a good deal by what people want."[105] Wittenmeier believed that the engineer's role was to give customers what they wanted, not to dictate the perfect environment.

But many engineers thought that the laboratory findings would help reeducate the public about the true conditions of comfort. One such believer in the power of science was Otto Armspach. Armspach was part of the engineering research establishment before he entered commercial air conditioning. While earning his engineering degree at Armour Institute of Technology, he had studied dust meters for the Chicago Ventilation Division, a pioneer in quantitative ventilation standards. Following his graduation, he worked as an investigator at ASH&VE's research laboratory. He took this laboratory experience directly into commercial theater air conditioning, working for the E. Vernon Hill Company by 1925 and then for CEC's Theater Division by 1929.[106] In 1930 he accompanied his article on theater air conditioning with a comfort chart issued by the Harvard School of Public Health. He predicted that patrons would soon ask theater management "whether they are prepared to control your rate of heat and moisture loss while you are in the theater and whether dust or other impurities will be prevented from entering your respiratory system in excessive amounts. These are the principal factors which determine your health and comfort."[107]

The ASH&VE laboratory research gave engineers confidence that their definition of comfort was more scientific and authoritative than individual notions of comfort. One engineering journal highlighted the importance of research when it noted that "along with this 'air conditioning consciousness' of the public—in fact, ahead of it—have gone the engineering research and experience which have made it possible to supply the condition which has created this acceptance."[108] One engineer acknowledged the value of the laboratory research when he noted in 1930 that for years now "sufficient information has been available as to comfort to enable one to know just what is required in order to obtain it . . . Comfort research has brought a new order into being."[109] With both science and experience at its back, CEC scoffed at "the absurdity of the familiar signs declaring that it is '20 degrees cooler inside,' " as a misrepresentation of comfortable conditions.[110]

Not surprisingly, air-conditioning engineers blasted the influence of theater managers who reinforced ideas about the primacy of cool temperatures. A disgusted air-conditioning engineer reported that one manager, preoccupied with the temperature, constantly fiddled with the thermostatic controls: "He wants to make his patrons as cool as he can without causing complaints," the engineer observed, "and will cool down until he gets a few complaints that it is too cool. He will then perhaps raise the temperature a little."[111] Carrier gave his colleagues a picture of the struggle between engineering expertise and the demands of the clients of these custom-made installations when he admitted that "the theater owner is not satisfied with conditions that perhaps theoretically may be best balanced. He insists on your giving him dry bulb temperatures that are slightly under 80 degrees, or perhaps 80 degrees as a maximum." Carrier was hopeful that as engineers discovered the best conditions for comfort, they could overcome exhibitors' insistence on low temperatures. "We are trying to educate the theatre owners to the importance of the control of relative humidity," he wrote, "and that the higher the temperature the more essential it is to have a low relative humidity."[112] Another engineer observed that "the public is being gradually educated to air cooling through a more restricted term: air conditioning."[113] But air-conditioning specialist Bloom was less sanguine: "I don't think we will be able to work against the psychology of the owner and the operator to carry quite as high a temperature as is advocated," he concluded.[114]

Engineers' writings were filled with complaints about exhibitors who pandered to public prejudice by focusing upon low temperatures. In 1926 Samuel Bloom recalled that "in the early years of air conditioning almost everyone who wanted to do any cooling in a theatre or elsewhere wanted to maintain 70 degrees and 50% humidity in the summer."[115] Indeed, the State-Lake Theatre in 1919 had advertised its system by boasting, "Wonderful Cooling Plant, Theatre Always Cool, 70 Degrees."[116] Such claims were often met by competitors, and Carrier and Lewis noted that once the rivalry over air-conditioning systems had begun, "the managers of other theatres have insisted upon cooling to severely low temperatures." They recorded disapprovingly that "theatre management . . . has found it necessary forcibly to prove the possession of a cooling system."[117] E. T. Lyle of CEC complained that the Roxy Theatre in New York City was just such a client. The theater maintained a temperature of 72–73 degrees at the insistence of the manager, whom Lyle characterized as "one of the class who think that they have got to freeze the people out in order for them to know that there is a cooling system."[118] While engineers struggled to balance temperature and humidity, theater mnagers focused instead only on temperature. A consulting engineer similarly recorded his frustration: "There is one peculiarity about air conditioning in the theater industry. To the theater people, it has always been known as cooling."[119]

Theater managers did what they thought would attract patronage and promoted an independent conception of comfort; in this way they both followed and led the public. On the most basic level, exhibitors installed air conditioning because the public liked it. "The theatre owners found that a comfortable theatre was a powerful drawing card," recalled one engineer about the rapid spread of air conditioning in the 1920s.[120] Since the general public was as yet unlikely to encounter air conditioning anywhere else, theaters provided their first formative experience with the technology. The pioneering role of motion-picture theaters in introducing the new technology to the general public is unquestionable. One architect recalled that "theatres were among the first commercial establishments to adopt air-conditioning extensively. Nearly everyone remembers the advertising that accompanied these early installations: '20 Degrees Cooler Inside,' 'Never Over 70 Degrees,' 'Arctic Breezes,' and 'Siberian Zephyrs.'"[121] Theater owners advertised the advantages of attending an air-conditioned theater by decorating the outside of their theaters with the iconography of icicles, icebergs, frost, and snow. Many theaters were designed with a system that provided a positive pressure in the auditorium, forcing cold air out through the lobby and onto the street to tempt passers-by. Engineers called this "advertising air." The public was getting its education in comfort from theater managers, not engineers or researchers.

The public's lack of appreciation for the ability of air conditioning to bring comfort to the theater through control of humidity as well as temperature persisted beyond the early years. As late as 1930, consulting engineer Malcolm Tomlinson called upon engineers to better educate consumers about the benefits of air conditioning's full range of control over temperature, humidity, and air movement. He complained that systems were operated at too low a temperature and remarked with considerable condescension that "of course, the general public is not yet in a position to understand the true significance of 'never over 70 degrees.' " He calculated that at a temperature of 70 degrees, combined with the usual relative humidity supplied to theaters and 400 feet per minute of air movement, the theater would be 8–11 degrees below the temperature needed for "perfect comfort." Tomlinson argued that "the cooling effect of conditioned air has been over-emphasized while the remarkable effects to be had from artificial weather, in comfort, have been neglected."[122] By concentrating so exclusively on temperature, this engineer argued, the public failed to see that the control of humidity as well as temperature, cleanliness, and air motion constituted the essentials of the natural climate. The recognition of the importance of humidity in comfort was a prerequisite to accepting air conditioning as artificial weather.

Despite the growing pile of laboratory reports, however, the public was still noncommittal about the importance of humidity to comfort and the superiority of man-made weather. The persistence of engineers' complaints from 1917 through 1930—that the owners of air-conditioning systems focused upon temperature rather than humidity as the key to comfort—suggests that many people failed to accept laboratory-defined standards. While Carrier wanted to install systems that carried 80 degrees and 55 percent humidity, owners were doubtful that a low humidity could really make 80 degrees feel cool. Carrier was optimistic that engineering standards of comfort would be increasingly accepted by the public, particularly if the model of air conditioning as man-made weather could convince consumers of the importance of humidity control.

COMFORT AIR CONDITIONING AND RECIRCULATION

Both CEC and the comfort air-conditioning industry in general were pleased that professional engineers unanimously rejected the term *fresh air*. While by-pass circulation was a tidy technical solution to the problem of humidity control at an economical price, it failed to comply with most of the ventilation ordinances then on the books. It could easily deliver the required 25–30 cfm per person, but only 15 percent to 25 percent of that air was "fresh" outside air. In effect, air-conditioning engineers who used bypass circulation were furnishing about the same amount of outside air that open-window advocates promoted.[123] Many ordinances were silent on the issue of recirculation, but a few, like the influential Chicago code, required that all recirculated air be sent through the air washer and purified.[124] That chilling of the recirculated air destroyed its usefulness for humidity control. Ironically, although the ventilating engineers as a whole had been anxious to maintain the 25–30 cfm standard as a way of mandating mechanical systems, comfort air-conditioning specialists saw it as a stumbling block. CEC engineer Lewis complained that "the established ventilation laws requiring a continual supply of 30 cfm per person are in direct conflict with the economy of first cost and operation of a dehumidifying system."[125]

As public health authorities adopted Dr. Leonard Hill's physical theory of air, they might have been expected to modify the old volumetric standards based on chemical vitiation. As early as 1918 the Chicago Department of Health's Ventilation Division frankly acknowledged that the theory of chemical vitiation of air was outmoded and accepted Hill's physical theory of air instead.[126] One consequence of the Ventilation Division's move toward implementing the new theory was the adoption that year of a "synthetic air chart" on which inspectors recorded the physical properties of the air in Chicago theaters.[127] As far as Chicago ventilation authorities were concerned, by 1918 the line between health and comfort was exceedingly thin. A comfortable theater was a healthful theater.[128] Good ventilation based on the findings of the most modern research was regarded as the best preventive against unhealthy conditions. Yet the department clung stubbornly to the 25-cfm standard. In 1922 the ventilation code still contained provisions associated with the older chemical theory of air quality, such as the 25-cfm requirement and the limitation of carbon dioxide to 10 parts per 10,000.[129]

The volume of air was not as crucial as the requirement that the air be drawn entirely from outside. This insistence on "fresh air" prohibited recirculation. E. Vernon Hill, a former inspector with the Ventilation Division, left to establish a commercial air-conditioning firm and by 1925 was sharply critical of Chicago's code. "We are not allowed by Health Department officials to recirculate air in theatres or public buildings, even though it is generally recognized that recirculation is desirable from every standpoint," he complained.[130] The vote of ASH&VE to delete the term *fresh air* from official usage found sympathetic support among air-conditioning engineers who argued that conditioned air from inside the theater that had been tempered and washed was far more wholesome than the raw air pulled into the system from the street. But the substitution of the term *outside air* did little to relieve their predicament.

In the end, the response of most air-conditioning engineers to ventilation ordinances was avoidance. One Chicago engineer admitted that "for years, air conditioning engineers have been hampered by ordinances requiring certain minimum quantities of air to be delivered into occupied spaces, 30 cu. ft. per min. per person being common. Necessarily, they are being disregarded today just as flagrantly and brazenly as the prohibition laws."[131]

MAN-MADE WEATHER

The inclusion of air conditioning in the motion-picture palaces created a definition of human comfort in the marketplace that paralleled the drive toward laboratory-defined standards. Like the advocates of public school ventilation, engineers who specialized in theater installations began with the attempt to equal nature, which soon yielded to efforts to create an indoor climate superior to the outdoors. Their success is illustrated by the public's use of air-conditioned theaters as a respite from the summer heat. One writer pointed out that "in extremes of weather, people go to the shows largely for the relief they hope to get from the outside weather."[132] Their determination to surpass the natural climate put these engineers squarely behind the conceptualization of air conditioning as man-made weather.

The analogy between air conditioning and climate had picked up its original support from manufacturers who wanted to transcend the limitations of geographical location and the handicaps of seasonal weather. But by the early 1920s the concept of man-made weather incorporated quantitative standards. The generation of these standards at the ASH&VE laboratory and at the Harvard School of Public Health did two things: it defined human comfort and also, by implication, an ideal climate that provided the conditions for human comfort.

Heating and ventilating engineers as a group supported quantitative standards because they protected the industry from professional challenges by other groups that targeted mechanical ventilation as one of the root causes of America's urban problems. But air-conditioning engineers supported quantitative standards for different reasons. They were enthusiastic about the experimental work of the laboratory because it incorporated the physical theory of air. Dr. Hill's interest in the physical properties of air positioned air conditioning at the center of the evolving profession, for the new technology promised to control just those aspects of the indoor atmosphere: temperature, humidity, cleanliness, and distribution of air.

The theater specialists in particular found the laboratory findings important because it addressed one of the major problems of their work: making theater patrons truly comfortable. Air-conditioning engineers found a natural enthusiasm among motion-picture exhibitors for transforming the theater from the worst place to spend a summer afternoon to the best place. In the beginning it was enough to make the theater more comfortable than the rival air-dome or other open-air entertainments. Comfort was defined comparatively. But laboratory research gave comfort quantitative limits and made humidity an integral part of its definition.

Thus the idea of air conditioning as man-made weather quickly translated an ideology of perfection into a specific set of hardware. The centrality of humidity control to climate, comfort, and air conditioning strengthened the dominance of the spray chamber as the preferred method of precise humidity control and the bypass system of circulation as the only economical way to achieve that precision. Thus, the ideals of the engineering profession were literally built into the systems that they sold.

Mass Production,
the Residential Market, and the
Window Air Conditioner, 1928–1940

*U*ntil 1930 factories and theaters provided the most important markets for the growing air-conditioning industry, for air-conditioning firms found these two types of clients eager to exploit the technology's potential for artificial climate. Both were vulnerable to seasonal restrictions that made the independence achieved with man-made weather especially appealing. In addition to their struggle with the natural climate, both created their own adverse climate within the building. Factories suffered from the intense heat generated by high-speed machinery and chemical processes, while theaters struggled with the body heat and smells emanating from densely packed crowds. They presented to air-conditioning firms a special type of interior space whose needs escalated with their increasing scale.

The separation from the natural climate that was a prerequisite of air-conditioning systems was congenial to both types of client. From its earliest days, the factory had frequently been seen as a world apart. While closing the windows made a real difference in the work life within, it was also symbolic of the gulf between work and leisure—between structured and free, economic and affective, natural and mechanical life. Manufacturers were not averse to coupling the controlled work life of the factory with an atmospherically controlled building. Indeed, the atmospherically controlled environment was another of the technical innovations in the factory, like the assembly line, that transformed the character of work and the autonomy of workers.

Motion-picture exhibitors also acquiesced to the separation of the theater from the outside world. They strove to create a retreat from the everyday world, one that was more luxurious, more exotic, and also more comfortable. That different environment was sometimes patterned on sumptuous European palaces. But the conceit of creating an artificial world within the walls of the theater was epitomized by "atmospheric" theaters. These theaters created the illusion that the audience was sitting outside under an open sky. Such a combination of the exotic and the natural is displayed in the main auditorium of the Atlanta Fox Theater, which surrounds the patron with a walled city and a pedestrian bridge across the proscenium, and Moorish tents suspended from the ceiling across the rear. In between was the night sky, complete with electric "stars," moving clouds, and the ability to simulate sunrise and sunset. Architects John Eberson and Charles Lamb were two of the most prominent practitioners of this style. Engineers gave architects and exhibitors an artificial environment to match their artificial world. While these client-company relationships were not without tensions over proper design and operation of the equipment, manufacturers and exhibitors provided the most complete acceptance of the technical community's ideal of man-made weather and the patronage to build these systems.

Yet custom-designed systems that placed environmental performance at the top of the list of priorities and retained the most legitimate pretensions to the claim of artificial climate, soon came to constitute only a small part of the air-conditioning industry. With the introduction of mass production to the industry, everything changed. Engineering firms slipped from the top of the industry, pressured by competition from manufacturing companies, and with that loss of status went the power and skills of the designer. Competitive strategy now centered far more on patents than on engineering skill, on costs rather than on performance, on equipment guarantees rather than on atmospheric guarantees. The vision of a rational, technological, and technocratic world inherent in the ideal of man-made weather dissolved in a characteristically American preoccupation with a democracy of goods achieved through mass production.

Nothing else captures that change so well as the appearance of the window air conditioner. Freed of ductwork and sewer connections, the window air conditioner also was liberated from its connection to the building itself. Its most important link became the one to the consumer, and the cost of ownership its most important feature. Unlike the custom-designed systems, the diminutive window air conditioner gave the consumer complete control if little power.

CUSTOM DESIGN AND THE RESIDENTIAL MARKET

Founding partner L. L. Lewis recalled that Carrier Engineering Corporation "went on our comfort-cooling drunk in '22," following the company's pioneering work in the Metropolitan Theatre.[1] The firm's success in theater air conditioning was built upon a solid base of engineering skill acquired in a decade of specialization in process air conditioning. Lewis estimated that by 1918 he had designed or directed the installation of 400 air-conditioning plants, worth an estimated $2.5 million, in more than 70 different industries.[2] Theaters were custom-designed installations, dependent upon engineering expertise. *Fortune* magazine explained to its readers that "air conditioning was born of, and grew up on, engineering skill. Man-made indoor weather, carried to the point of painstaking precision, is strictly a twentieth century accomplishment—and the engineers did it."[3] With a strong patent position to bolster its de facto leadership, it is no wonder that CEC felt heady over the prospects now open to it.

Yet the company's enthusiasm for comfort air conditioning inexorably took it far beyond its traditional base. The vast, untapped market of residential air conditioning beckoned just beyond the motion-picture theater, and the company's interest in it increased as industrial demand slowed due to the Depression. Willis Carrier had no doubt about the direction of expansion and predicted in 1931 that "the greatest market of all will, of course, be the home."[4] However, home air conditioning presented a whole new set of technical problems with little relation to the old custom-designed systems. In the years from 1928 to 1940 the character of air conditioning changed from a system that extended throughout the building to an appliance set in the window. This change, pushed by the drive to make air conditioning affordable for the individual consumer, opened up the industry to new companies, whose previous experience in small refrigeration units, mass production, or merchandising compensated for their lack of sophisticated technical expertise in designing systems for climate control.

In the drive to adapt large commercial systems to the home market, air conditioning underwent considerable technical change, becoming smaller, simpler, and safer in operation. In addition, a host of partial systems that lacked the technical capabilities of year-round environmental control proliferated in response to the potential of the residential market. By 1930 the conflict over the design of comfort air conditioning had been resolved in favor of a package of ideal standards, precise humidity control, and substantial recirculation that privileged the bypass patents and the Auditorium Conditioning Corporation. The company benefited from the conjunction of experimental science, engineering ideology, advertising, patents, and economic concentration. But humidity control, which dominated the engineering community's understanding of air conditioning, had only a weak appeal to consumers who now gladly traded true man-made weather for lower prices.

Residential air conditioning was not strictly new in 1928. The earliest installations may be Alfred Wolff's designs for favored clients that incorporated winter humidity control, but these systems apparently provided no summer cooling. In addition, in 1913 Carrier had designed and installed an elaborate system for the $900,000 mansion of Charles Gates in Minneapolis, Minnesota, which included a $2,200 humidifier and a $2,900 refrigeration plant.[5] Unfortunately, the owners died before the house was finished, so that it was never occupied and the system was probably never used. Such an installation was not much different in size, cost, or equipment than air-conditioning systems found in theaters.

Just as process air conditioning in industry provided a poor model for commercial comfort air conditioning in theaters, so commercial air conditioning proved to be a poor model for residential air conditioning. While the aim of both was comfort, the equipment used by theaters was bulky, often posed a safety hazard, and required a licensed engineer for operation.[6] One of the chief technical problems in adapting commercial machinery to residential use was the heavy reliance of commercial installations upon the air washer. Air washers both cleaned the air and controlled humidity, thus providing two of the four essential features of air conditioning. They were often used for cooling as well, using either cold well water or refrigerated water in the spray chamber. The advantage of the air washer over cooling coils was that it did not rust; the spray itself provided a condensing surface for excess humidity. In addition, it eliminated the problem of ice build-up on cooling coils, which severely restricted the flow of air through fan-coil systems. Most important, the use of air washers or spray chambers permitted precise humidity control. However, air washers were bulky and impractical for most residential applications. Clearly, an alternative would be needed.

In addition, commercial comfort air conditioning did not provide a suitable solution to the dangers of refrigerants, most of which were both toxic and inflammable. Beginning in 1915 the New York City Fire Department regulated refrigeration installations to ensure public safety, and a number of other cities followed its example. Of the four commonly used refrigerants in cooling equipment in 1915—ammonia, carbon dioxide, ethyl chloride, and sulphur dioxide—carbon dioxide was the safest. In recognition of the toxicity of refrigerant gases, most theater installations used carbon dioxide as a refrigerant, for it could rise to as much as 10–15 percent of the atmosphere without danger, while fractional percentages of other refrigerants could cause death.[7] However, carbon dioxide required higher operating pressures than other refrigerants, and the machinery to handle it was consequently bulky. Like the air washer, the carbon dioxide refrigerating machine was too large to be practicable for residential air conditioning.

Carrier seemingly had found an alternative to bulky carbon dioxide systems when he applied for patents on a centrifugal refrigerating machine in 1921. The next year CEC began to manufacture the centrifugal machine using dichloroethylene as a refrigerant, and it proved to be important in the company's theater installations. Lewis regretted the use of ammonia in the Metropolitan Theatre and resolved never to install another system like it.[8] Indeed, CEC installed the new centrifugal technology in its next comfort system, for a pair of Texas theaters, and followed that with a system in the highly visible Rivoli Theatre in New York City in 1924. In 1926 Carrier adopted a new refrigerant, methylene chloride, for the centrifugal machine, thereby providing greater cooling capacity without redesigning the machine.[9] The centrifugal machine was much less bulky than its carbon dioxide predecessors, but it was not designed for small cooling loads. All these improvements made large comfort systems more efficient, but did nothing to help CEC move into the residential market.

THE WEATHERMAKER AND THE PERSISTENCE
OF ENGINEERING PERFECTION

In 1928 CEC designed its first small machine for the residential market in 1928. Called the Weathermaker, it was a gas-fired heating system that included forced circulation, cleansing, and humidification. It offered complete environmental control, but only for half the year because it did not incorporate cooling. Such equipment, ambitious in one respect and limited in another, was called a "winter air conditioning system." A prototype was installed in the homes of several CEC employees in 1927, and the company began marketing the Weathermaker the next year.

A subsidiary, the Carrier-Lyle Corporation, was organized in June 1928 to handle this new kind of equipment, and a certificate of incorporation was filed in August.[10] The company recruited Cornelius R. Lyle, J. Irvine's brother, from Armstrong Cork Company to be executive vice president, and hired Vincent S. Day, a leading researcher in warm-air heating at the University of Illinois mechanical engineering department, as chief engineer.[11] In addition, it hired Margaret Ingels, who had been supervising the New York Ventilation Commission's test in Syracuse public schools.[12]

The air-conditioning industry was forged by, and reflected the values of, an all-male engineering fraternity. Ingels is a notable exception. She graduated from the University of Kentucky in 1916 with a degree in mechanical engineering, at a time when only sixteen other women in the United States had earned an engineering degree. Her work at the ASH&VE laboratory made her one of the pioneers of the new technology. Her employment at Carrier-Lyle represented a very different turn in her career. Unlike her early work as first an estimating engineer and then a researcher, her work at Carrier-Lyle centered on public relations rather than the technical aspects of engineering. Lewis recalled that "the reasoning was that inasmuch as the housewife would have an important voice in that kind of an investment, a woman engineer could be very effective."[13] Because she was a woman in engineering, she was viewed as doubly qualified—knowledgeable about air conditioning and about women consumers as well.

Despite CEC's recognition that the new market required a very different technology, the company's first residential product reflected older traditions. The Weathermaker embodied CEC's long and fierce commitment to air conditioning as a comprehensive system of environmental control including humidification. After Day's redesign of the prototype, the machine included several ambitious features. Rejecting an intermittent heating cycle, he wanted a machine that supplied heat continuously in varying amounts depending upon the need. To meet that requirement, a multistep electrical control for the gas had to be developed as there was none on the market. The new control was remarkably successful, but very costly. In addition, the Weathermaker was designed to provide humidification independent of the heating system, but the small amount of heat necessary to maintain perfect temperature did not produce adequate evaporation. As a result, the machine was equipped with an independent system for maintaining constant humidification. This, too, was costly. The result was an innovative, expensive machine. Both these characteristics explain its ultimate failure in the market.

The company advertised its product with the promise that "now you can have made-to-order weather in your home."[14] The price was about $3,000.[15] Surprisingly, the company found customers in that high-end market. By the end of 1928 the East Orange District had sold twelve models, and by the end of 1929 it had sold sixty-eight. In all, Carrier-Lyle sold approximately one thousand Weathermaker machines.[16] The roster of customers, one engineer recalled, "read like a page out of a Blue Book."[17] Initial customer reaction was encouraging; a Long Island customer wrote to the firm to say that his machine "has worked perfectly since it was started in November 1929. I believe it is the best heating equipment in every way for this part of the Country."[18] Another took the company's advertising rhetoric to heart, writing, "You have made it possible for me to have June in my home in January and I wish to express my great appreciation."[19]

This shift to a new market with its own small-scale equipment accelerated CEC's transition from an engineering company to a manufacturing firm. The first move in this direction had occurred in 1922 when the company bought a building at 750 Frelinghuysen Avenue in Newark, New Jersey, for the manufacture of Carrier's centrifugal refrigerating machine.[20] Then in 1928 it acquired a second, 112,000-square-foot building at 850 Frelinghuysen (formerly occupied by the Edison Lamp Works) to manufacture the Weathermaker.[21]

The transition from engineering to manufacturing proved rocky. The Weathermaker was plagued with technical problems, some of which could be traced to the manufacturing facility. The most serious and persistent was leakage in the heat exchanger, which was fabricated out of a new kind of sheet stainless steel. The factory failed to anneal the welds in the heat exchanger properly, and depending upon the quality of the gas burned in the unit, the exchanger developed cracks sometimes after only a few weeks of service. As *Fortune* recounted this episode in the company's history, the Weathermaker "developed a distressing tendency to leak on ladies' rugs."[22] Day recalled that "there were many small complaints, some not so small . . . Thus Carrier-Lyle had a lot of half satisfied customers."[23] It was the worst possible scenario: Carrier-Lyle had a thousand of the nation's most influential people as customers, and many were extremely unhappy with the company's product. The company was quick to replace the defective units, but by the end of 1932 it had lost $250,000.[24]

Trouble with the existing model was matched by the near complete collapse of sales after the October 1929 stock market crash. Especially in the Northeast, homeowners had hailed the gas-fired Weathermakers for liberating them from the mess and inconvenience of coal. Even though gas heating was estimated to cost two and a half times as much as coal, few customers regretted the extra expenditure—at least until the economic downturn. By 1930 some owners were inquiring into methods for converting their Weathermakers to less expensive oil, and the company's attempts to accommodate them produced only "sooty messes." Carrier-Lyle was dissolved in 1936, and the replacement program was shut down in 1937.[25]

DEPRESSION-ERA ECONOMICS AND IDEOLOGY

The Depression changed everything about the move into residential air conditioning. If the home market in the late 1920s had called for machinery that was small and inexpensive, the years after 1929 only intensified the importance of those characteristics. No longer could a company build perfectly engineered but expensive equipment if it had any intention of reaching a mass market. But, surprisingly, few of those in the air-conditioning industry thought that the Depression was an unsurmountable barrier to its expansion. Indeed, for those who had faith in technical progress, new products seemed exactly what was needed to spur a sluggish economy out of crisis.

The onset of the Depression not only made markets tighter, it also presented an ideological challenge to technological development in general. As the economy plunged deeper, critics argued that the economic collapse had been partially caused by overproduction and a runaway technology that increased unemployment by replacing men with machines. In England there was talk of a moratorium on research and development to allow society's moral and social development to catch up with recent scientific and technological growth.[26]

But few in the engineering community felt that technology per se was in any way responsible. The most radical instead blamed capitalists' control of productive technology. The engineer Henry L. Gant had developed just such a critique of the uses of technology in American society following World War I. Struck by the extent to which factory owners used their machinery for profit rather than maximum "efficiency," Gant had articulated a technocratic policy that called for the management of production by technical experts who put rationality and productive gains above mere profits. The call for a more technocratic leadership made a brief, fiery appearance on the American national scene in response to Depression conditions, but soon fizzled. The most politically charged versions of technocracy seem to have had limited appeal among both engineers and the general public during the 1930s. Perhaps Herbert Hoover's election as the country's first engineer-president blighted technocracy's appeal from the beginning.

Although most Americans were reluctant to believe that engineers could provide the leadership needed to bring the nation out of its dilemma, many believed more blindly that technical progress itself would heal the economy. Certainly, the leaders of individual companies and whole industries felt that only more technology would boost sluggish sales. The air-conditioning industry was no exception. Willis Carrier told the National Association of Power Engineers that he still had faith in technological development. "At this time, we again hear the cry of those who would place upon the 'machine age' the responsibility for unemployment and economic instability. The 'gloomy Dean' Inge of England has gone so far as to beg scientists to take a ten-year holiday that civilization might catch up," Carrier said. However, he saw technical progress as inevitable. "We have no further choice in the matter. Man *will* go on. New devices create new wealth; call for *new* raw materials and create new uses for old ones . . . As an engineer, I have only confidence in the future to which liberating mechanisms are leading us." Carrier endorsed the notion that the best thing for the economy was more technological development, not less: "The immediate situation imposes a crying need for new industries, which will contribute wealth and employment in a degree comparable to that produced by the automobile, the radio, and the household refrigerator."[27]

The belief in the power of new technology to create wealth was not theoretical, but rather was based upon recent history. Engineers alluded again and again to America's experience with the radio. Its history spelled out for the doubtful a success story of invention, mass production, and popular marketing, creating national wealth that assured the country of entrepreneurial rewards, full employment, and consumer luxuries. The history of the radio suggested the possibilities for the new home air conditioner.

Indeed, such sentiments were not simply the view of a single individual or company. If technology in general could create the national wealth necessary to solve America's economic woes, old-line heating and ventilating companies and traditional refrigeration firms believed that air conditioning in particular could be a powerful force for stimulating the economy and helping to end the Depression. *Refrigerating Engineering* noted the "optimism that refrigeration 'would pull us out of the depression.' "[28] The editorial pages of *Heating, Piping and Air Conditioning* retailed the belief that air conditioning "can directly or indirectly give millions of people jobs and go a long way toward restoring employment to normal conditions."[29] General Electric hinted that air conditioning "may be the industry which will take the place of the automotive industry of 1922 in bringing back prosperity."[30]

Industrial analysts were optimistic about the future of air conditioning, and their enthusiasm was quickly communicated to the larger public. For example, Westinghouse saluted the contributions of the heating and ventilating industry in April 1930 over a nationwide network of thirty-one broadcast stations, a tribute featuring a forty-piece orchestra and large chorus.[31] Such promotional efforts and optimistic predictions prompted an editorial in the *New York Sun* that told readers, "It won't be long now. When the first 'talkies' were privately exhibited, the Hollywood barons said optimistically that in two years the whole country would be seeing them. It did not take six months. Demand spurred ingenuity and production. So it will be with general air-conditioning."[32] However, when *Good Housekeeping* and *Better Homes and Gardens* predicted that the better class of homes would have year-round temperature and humidity control within a short time, *Refrigerating Engineering* confessed that "the engineering fraternity will hope to agree, but may as well say now, that as a group, it does not know in general how it is to be done."[33] To many engineers, residential air conditioning with cooling seemed futuristic rather than immediately practical. Indeed, most engineers felt that considerable technical innovation would be necessary to make domestic air conditioning feasible.[34] The predictions seemed to have a driving force of their own, however. The business community's belief in the redeeming power of new technology and its own desperate plight spurred technical innovation.

Some manufacturers, particularly from the refrigeration industry, challenged the dominance and the exclusiveness of the older engineering and heating and ventilating firms, such as Carrier Engineering Corporation. In the period when air conditioning was associated with the precise control of humidity in industry, few refrigeration specialists had been willing to challenge the heating and ventilating engineers' control of the market. David L. Fiske, the editor of *Refrigerating Engineering*, acknowledged that "the art of air conditioning originated in the art of ventilation, which was an art understood by a very small number of men and manufacturers. The art has continued to be 'closely held.' "[35] Refrigerating engineer Everett Ryan pinpointed informal knowledge—what heating and ventilating engineers called "the art of engineering"—as the greatest barrier to entry into this field: "The very crux of air conditioning is air movement, and practically all the crucial knowledge of it in the hands of a few experienced men. Rules and theories have never taken the place of their experience and horse sense."[36] As late as 1931 Fiske acknowledged that "air conditioning was distinctly a contracting, rather than a merchandising affair. It was thought of in units of a hundred tons, and one or two engineering outfits did all the work."[37] As it became increasingly clear that the average homeowner could not afford the expensive tailor-made installation that was their specialty, Ryan pointed out that "the ventilation fraternity, have never taken the comfort cooling job, especially on the small scale, at all seriously. They have found markets in theatres and public assemblies, to be sure, but they are not the ones who talk of cooling the homes of America."[38]

Increasingly, refrigeration companies targeted air conditioning as an important market. One industry journal was convinced that "domestic air conditioning ... has an excellent chance of being refrigeration's greatest future market."[39] Fiske reported that the refrigeration industry's interest in air conditioning was "a result of the changed economic situation," particularly "the lifelessness of the capital goods industry and the heavy machinery business."[40] "In such periods," he claimed, "goods sold to the layman tend to be more active and profitable than those sold to industry. Thus, everyone rushed into air conditioning."[41] Indeed, Fiske noted the effect of the Depression in pushing established companies to adopt new products: "Companies which had been ultra conservative have saved their skins only by shifting to new and modern lines of goods, which they would never have considered otherwise."[42] The popularity of theater air conditioning placed refrigeration skills at the center of any future developments in comfort air conditioning. In the public mind at least, comfort air conditioning meant cooling.

DOMESTIC TECHNOLOGY AND THE REFRIGERATOR

Refrigeration companies were not all equally successful in tapping the air-conditioning market, despite the importance of cooling to comfort air conditioning, for not all companies possessed the appropriate technology. The carbon dioxide refrigeration machines of the theater installations were poorly suited to the space and pocketbooks of residential consumers. But refrigeration companies that had pioneered small-scale refrigeration for the domestic refrigerator established an early lead in residential air conditioning. Just as the example of radio served to inspire a daunted and depressed industry to embrace new technology, the domestic refrigerator became the practical model for the development of small comfort air-conditioning systems.

In many ways, the refrigerator provided solutions to the problems of mass-producing small equipment that was safe and simple to operate. Domestic refrigerators were offered on the market as early as the 1890s but were not widely adopted until the 1920s. In that developmental period, engineers sought technical solutions to the problems of small-scale design and domestic operation.

One of the first refrigeration machines designed for domestic use was the Audiffren-Singrun machine. It originated with the work of Marcel Audiffren of Grasse, France, who received a U.S. patent in 1895.[43] *Ice and Refrigeration* reported that the early model, produced in 1904, was "crude," but Audiffren and Albert Singrun filed for improvements in 1905, and the journal approvingly noted the "perfected" model in 1908.[44] A group of Americans secured the domestic patent rights and formed the American Audiffren Refrigerating Machine Company. General Electric (GE) agreed to manufacture the equipment, and Johns-Manville Company contracted to market it. GE's Fort Wayne works turned out the first machine in 1911 and produced 150–200 a year until 1928.[45]

From the first, the Audiffren Refrigerating Machine Company had its eye on the domestic market, at least the upper-class domestic market that overlapped with small-scale commercial concerns. The machine was manufactured in four sizes: ⅕-ton, ½-ton, 1 ton, and 2 tons. The smallest model made 11 pounds of ice an hour; owners also had the option of making a smaller amount of ice and maintaining a refrigerated food cabinet. If the refrigeration capacity was small and particularly appropriate for the home, the machinery was not. It was very large for most homes. Some of the Audiffren machines were a single bulky package, but other models were a "split system," with the refrigeration apparatus in a remote location (usually the basement) and connected by pipes to the refrigerated cabinet in the kitchen.[46] Operation of the machine was not automatic; users had to turn on the power and the cooling water for the condenser and shut them both off to stop the cooling process.

Manufacturers clearly worried that consumers were not competent to correctly operate a machine that contained a toxic gas under pressure. One engineer recalled that "when it was first proposed that a small, automatic refrigerating system be attached to the household refrigerator and operated without the attention of an engineer, a number of objections were raised, based on experience with large plants."[47] The manufacturer's response was to design into the machine the utmost simplicity and safety. In 1913 an Audiffren-Singrun brochure reassured customers that "absolutely no skill is required to run the Audiffren-Singrun Machine. Its construction is simplicity itself. 'Your cook can run it' . . . The operator doesn't have to understand its principles any more than he has to understand the principle of his watch or electric motor."[48] It emphasized that there were no connections to leak refrigerant, the pressure was not sufficiently high to cause an explosion even if the machine were abused, and any liberated refrigerant gas could not produce fatal or even dangerous results.

Not everyone was persuaded of the practicality of a domestic model. J. F. Nickerson, publisher of the pioneering trade journal *Ice and Refrigeration*, noted in 1915, "The domestic machine . . . has received a vast amount of attention but the problems confronting the use of such a machine are enormous, and while many of them have been solved, it cannot as yet be considered a commercial success for use in the average-sized family."[49] In 1917, at a meeting of the St. Louis branch of the American Society of Refrigerating Engineers (ASRE), one of the guests exhibited a small rotary machine using sulfur dioxide as a refrigerant, undoubtedly an Audiffren-Singrun machine. The disadvantages discussed among the engineers included the high initial cost of the machine for the average household, as well as the "liability of operating troubles with the available household help."[50] Despite the slogan "your cook can run it," these engineers were skeptical. They viewed housewives and immigrant domestic help as a particularly illogical and irrational segment of society. Even by 1917, then, the domestic refrigerator was not universally accepted.

In 1915 and 1916, as a result of its experience with the Audiffren machine, GE began to design a machine that was less complicated and less expensive. After producing a series of test models in 1918, GE developed a refrigerator with the motor enclosed in the compressor case, thus eliminating a perennially troublesome stuffing box between the motor shaft and the compressor. The "enclosed motor," as it was called, was an important improvement in refrigerator design. However, it was the appearance of the air-cooled condenser in 1923 that made the biggest difference in sales. The initial condenser was water-cooled, and the copper coil evaporator was submerged in a brine tank in the upper part of the cabinet; the air-cooled condenser eliminated the necessity of a sewer connection to dispose of the cooling water. Sales of household and small industrial compression units totaled 4,000 in 1921; by 1930 sales had reached 1 million. Arthur J. Wood gave much of the credit for such rapid expansion to the development of the air-cooled condenser.[51] In 1927 GE used these improvements in its immensely popular Monitor Top refrigerator.[52]

The Kelvinator Corporation was another company that began as a manufacturer of domestic electric refrigerators. Kelvinator originated in a partnership of two businessmen and an engineer. On 14 September 1914, broker A. H. Goss and purchasing agent E. J. Copeland, both of Detroit, contracted with Harvard-educated engineer Nathaniel B. Wales to develop his ideas for a practical domestic refrigerating unit. After rejecting anhydrous ammonia as a refrigerant because of its high operating pressures, its toxicity, and the cumbersome machinery it required, Wales, like Audiffren, settled on sulfur dioxide. Sulfur dioxide provided low working pressures and was comparatively nontoxic, nonflammable, and nonexplosive. The first test models lacked complete automatic control and exhibited problems with gas leakage, but by 1918 the company had sold sixty-seven units from a sales office in Detroit. Two years later 600 Kelvinator machines were in operation. Like the early Audiffren-Singrun and Brunswick machines, the Kelvinator was a "remote" type, or split system; while the food cabinet was located in the kitchen, the refrigeration machinery was placed in the basement, and the two were linked by a series of pipes that circulated the refrigerant. In 1925 the machinery was consolidated into one casing, and in 1927 Kelvinator opened a factory in Detroit that it claimed was "a marvel of 'straight line' production efficiency rivalled only by the automotive concerns."[53]

Thus by 1928 the refrigerator had evolved into a self-contained appliance. Thanks to the air-cooled condenser, it was independent of sewer connections; through the adoption of sulfur dioxide and the development of the enclosed motor, all the machinery was small and safe enough to fit into one room; and its operation was regulated by automatic controls. Such a machine, mass-produced and affordable, predicted the direction of air-conditioner development.

THE EARLY SPLIT-SYSTEM ROOM COOLERS, 1929–1932

The refrigerator manufacturers moved directly from the production and sale of refrigerators to residential air conditioning. They had considerable technical experience with small-scale refrigeration units, automatic control, and mass production. By late 1930 Kelvinator, Frigidaire, and GE had announced or planned to announce a small air conditioner.[54] All were designed to cool just one room. Such a modest machine would bring the costs down in two ways. First, it would be cheaper to buy and operate than the central air-conditioning stations that characterized theater installations. Second, single-room cooling virtually eliminated expensive runs of ductwork.

Frigidaire was the first company to offer a room cooler, reportedly introduced in 1929.[55] It was developed by Frigidaire's research engineering department, headed by Harry Blair Hull, after sales representatives had sold an improvised room cooler in the field for the preceding four or five years. The new cooler resembled early refrigerators more than contemporary models. It was a split-system design, with the sulfur dioxide condensing unit in the basement, connected to the cabinet by pipes. It was heavy and bulky. The room cooler measured 28 × 18 × 49 inches and weighed 200 pounds, while the remote condensing equipment weighed 400 pounds. Although it was considered suitable for the home, company engineers remembered that "quite a few commercial installations were made."[56] Small businesses seemed to like the lack of ductwork, for such customers as the Hollywood Night Club in New York City installed multiple coolers.[57] The 1930 model was designed to cool 300 square feet, and the company promised that the unit would lower temperatures about 10 degrees and humidity about 10 percent.[58]

General Electric began making room coolers at the same time. The company built a dozen in 1930 and made plans to produce another dozen in 1931. The GE room cooler also used sulfur dioxide refrigerant, and like the Frigidaire model, it was both bulky and expensive. Providing enough cooling capacity for a room of 3,500 cubic feet, the 560-pound machine itself took up a space that was slightly more than 30 inches wide, 49 inches high, and 24 inches deep.[59] It cost $950, not including the wiring and plumbing required to install it, and during 80-degree weather it could provide 10 hours of cooling for 38 cents per day.[60]

It was an unusual customer who spent so much on conditioning one room. One GE customer was Mrs. John L. Kellogg, an allergy sufferer. For eighteen years she had avoided hay fever season in the traditional manner—by seeking a more favorable natural climate. But with a new GE room cooler, she simply closed all the windows in her bedroom "and enjoyed more restful sleep than I have at any hay fever resort."[61] Indeed, GE itself may not have envisioned a large market for the product; the engineering department imagined its greatest usefulness might be as a "feeder." "A person having gotten the benefit of one cool room in his house should be a prospective for larger equipment to cool the whole house," a GE official speculated.[62]

Carrier-Lyle was not happy with its competitors' entrance into the market. Although GE claimed that excess humidity would condense on the finned coil evaporator, Carrier-Lyle thought this claim questionable and challenged GE. Carrier-Lyle held to a long company tradition and now established standard: if a product did not have humidity regulation, it was not an air conditioner. In the August heat of 1931 the company's advertising manager, Brewster Beach, was blunt in his comments to GE officials: "Perhaps calling this piece of equipment an 'Air Conditioner' is improper because it is not an air conditioner . . . More definitely we believe G. E. knows enough about air conditioning to realize this device is not an air conditioner."[63] The company was equally adamant that the Frigidaire product was a cooling device and not an air conditioner. Because it lacked an outside air intake, Carrier-Lyle protested, "it represents the distinction between a simple cooling unit and an air conditioning unit."[64] Carrier-Lyle Corporation insisted that the restrictive definition of air conditioning be applied in the marketplace.

Such a critical posture sprang from the fact that CEC was designing a small room cooler as well. Its experience with the Weathermaker had convinced the company that designing small-scale residential equipment was a difficult proposition and that it needed additional expertise. To that end, sometime before December 1930 talks began about a three-way merger between CEC, the Brunswick-Kroeschell Company, and the York Heating and Ventilating Company. Both Brunswick-Kroeschell and York had experience in developing small-scale equipment that would be useful for CEC's new room-cooler venture.

The idea appealed to Brunswick-Kroeschell executives, who worried that the Auditorium patents would choke off their comfort-cooling business. And Brunswick-Kroeschell's experience with small, self-contained water-cooled condensers and fully automatic controls appealed to CEC. Brunswick-Kroeschell produced small ammonia machines, called B-K Jrs., designed for small businesses. The company advertised them as "portable" since their unitary construction meant there was little ducting or piping left behind if the customer moved from rented quarters. In 1932 the B-K Jr. compressor was redesigned to better suit Carrier Corporation's room-cooler ambitions.[65]

Just as important was the expertise of York Heating and Ventilating, which had built a prototype unitary heater for factories as early as 1922. Before 1918 York had been a prosperous sheet-metal contracting firm that fabricated ducts for central heating and ventilating installations. When Thornton Lewis and Paul Gant bought into the firm that year, they brought with them the idea that the expensive ductwork in which the firm specialized was vulnerable to technical innovation. The substitution of several small, self-contained fan-coil heaters for the large central-station system would eliminate the need for ducts. They resolved to transform the business into a manufacturer of small, self-contained equipment. The company's research resulted in the development of a lightweight heating coil called Superfin, and a small propeller fan named Kroy, which was York spelled backward.

By 1926 York's leaders had applied their vision to self-contained cooling units as well. They established an air-conditioning department in November. Gant, the vice president of the new division, patented a cooling apparatus in 1928.[66] In addition, York hired CEC engineer R. R. Taliaferro to join its research and development department. Before long the company had three cold diffusers, of which two mimicked its heating design and one wetted the cooling coil with a brine spray. York's air conditioner was designed for industry, but CEC believed elements of the York equipment could be successfully adapted to smaller residential units.[67]

The merger was completed in December 1930.[68] To no one's surprise, the resulting company was named Carrier Corporation, after its new chairman "whose publicized name and fame are probably its greatest single asset."[69] Cooperative research was begun in Newark, and in 1931 Carrier Corporation announced the Atmospheric Cabinet. Like the Frigidaire model, it was a split system, with a fan, cooling coil, and filter connected to a remote refrigerating machine. Although both the cabinet and the refrigerating unit were preassembled in the factory, the remote installation meant that considerable labor and cost were involved in installing an Atmospheric: drain lines for condensate from the cabinet, refrigerant connections between the cabinet and compressor, water connection to the condenser, electrical connections to all. Installation costs ran $450–$750.[70] The first six units were installed with some modification on 25 May 1931 at Lehman Brothers' office in New York City.[71]

In February 1932 Carrier Corporation made the Atmospheric Cabinet (still a split system) commercially available. It listed for $880–$2,000, depending on size.[72] A company publicist in May, marking the appearance of the new small machines, noted that "the summer of 1932 will witness the initial widespread offering of these units . . . and by fall of 1932 we should know much about their potentialities."[73] But sales of the Atmospheric Cabinet never materialized. Indeed, production stopped altogether for a while, resuming in the 1936 season.[74]

Perhaps the Atmospheric Cabinet looked clumsy in comparison to its rivals. By 1932 there was more competition than ever from large refrigerator manufacturers. That year the awkward split systems began to disappear, leaving the Atmospheric Cabinet behind.

The use of new refrigerants that were especially safe gave impetus to the room cooler. Drawing upon its long experience with domestic refrigerators, GE insisted that it had no qualms about the safety of its 1930 sulfur dioxide model. One official argued that "since this is approximately the same gas charge used in our small domestic machines, and since we have had absolutely no trouble from leaks in the one million machines we have already sold, we do not feel that the use of sulphur dioxide in a sealed machine in these small quantities is dangerous."[75] In 1930 General Motors announced that chemists Thomas Midgley Jr. and Albert L. Henne had developed a new refrigerant, Freon, that was particularly safe. David Fiske noted that "almost invariably the availability of low pressure refrigerants, Freon and methyl chloride, has been a factor in the current attitude of hopefulness that this practice can be built up. Nobody thought of it years ago on account of the universal danger ascribed to the refrigerating machinery, limiting it to larger units under technical super-vision."[76]

In addition, rivals appeared ready to meet the challenge to produce a full-fledged air conditioner appropriate for one room of the house. Frigidaire offered three types of self-contained air conditioners. All of them boasted the full range of air-conditioning functions of cleaning, heating, and cooling the air and controlling its humidity.[77] Meanwhile GE established a separate air-conditioning department and was field-testing units that provided "all the functions for complete air conditioning."[78]

In 1933 GE brought out a room cooler that, unlike the model CEC had tested, was an all-season air conditioner. GE did it by linking up with American Radiator and the Campbell Metal Window Company. Each of the three companies found that its product offered only one part of a larger technology. For example, Campbell Metal Window had marketed a "soundproof" window that included a filtering and ventilating device to allow customers to open the window for fresh air without having to listen to objectionable urban noise. The new air conditioner combined the Campbell filter with the American Radiator cabinet heating and humidifying device, and GE's electric control and refrigeration mechanism.[79]

By the next year rival Kelvinator had twenty-four models ranging from ½ to 8 tons of capacity and including a self-contained unit as well as a split system. Westinghouse's two-year effort in development yielded a small self-contained unit as well.

FROM THE AIR-CONDITIONING SYSTEM
TO THE AIR CONDITIONER

The development of self-contained room coolers was a natural response to the needs of the market and the traditions of the refrigeration companies. Retailers also found the transformation of the air-conditioning system into an air-conditioning appliance a tidy solution to their own interests. Utility companies in particular pushed for a self-contained air conditioner that could be added to their stable of electrical appliances for the home.

During the Depression power companies saw air conditioning primarily as an opportunity to increase their residential energy load to substitute for lagging industrial markets. Noting that air conditioning used from two to four times as much electric current as a refrigerator, *Business Week* reported in November 1930 that power and light companies believed "that people will pay for adequate defense against such weather as we had last summer." They "confidently expect that there will develop a strong demand for them next year."[80]

When the National Electric Light Association (NELA) picked a symbolic "all-domestic" team for 1932, a charming picture of a football team accompanied its opinion that "for *tackles* on our all-electric team, the coach might well pick Health Appliances and Air-Conditioning to open up new holes . . . In them is amazing potential power for gaining the service goal."[81] Stimulating household consumption of electricity was not a new practice for most power companies. For years they had promoted and sold electric appliances to customers to boost power usage, including such items as fans, irons, waffle irons, and toasters. In 1931 they drew upon their experience in merchandising other appliances as a model for the promotion of air conditioning. One utility executive told his colleagues that "in the experience of every utility exists precedent that can be utilized in the promotion of air conditioning. We need cite only one familiar example—the electric refrigerator . . . The whole range of our experience with domestic appliances is applicable."[82]

The familiarity of the practice of promoting new appliances meant that utilities were quick to respond to the opportunity presented by residential air conditioning, and their rapid response put them in the forefront of the movement to promote the technology. Whether or not a utility company undertook to sell air conditioning equipment, utility executive C. E. Michel argued that "a sales job has to be done," and "no satisfactory substitute has been found for the utility."[83] Reasoning that the most persuasive sales technique was experiencing the comfort of air conditioning, companies like Philadelphia Electric in 1931 and Detroit Edison Company in 1932 air-conditioned their own offices to promote comfort systems.[84] "A hot wave sells more room coolers than all the logic of the industry," one magazine reported.[85] Once interested, most potential customers still wanted to know how the machinery worked, and sales representatives with a technical understanding enjoyed the most success. Customers then weighed "the cost against the benefits of personal comfort and health."[86]

The utilities' analogy between air conditioning and other domestic appliances was not a precise match. Air conditioning had long been a mechanical system, composed of bulky air washers or cooling coils, fans, compressors, and motor, together with ductwork that spread throughout a building. Because of its two-part nature, air conditioning was not physically discrete like other domestic appliances. It was not simple to install, either, for the water-cooled machinery required a water hook-up and a sewer connection. It was not considered portable but instead remained part of the house when the owners moved.

As a consequence, the utility companies were strong proponents of the transformation of air conditioning into the air conditioner—a plug-in appliance that was factory-produced, portable, and required no installation other than plugging into the standard household electrical outlet. Although engineers concerned with technical performance disliked the limited capacity and poor atmospheric control of the room and window air conditioner, *Refrigerating Engineering* noted that the room cooler meets "the demands of the 'just plug it in' school of salesmen."[87]

So eager were the power companies for new loads that some displayed a decided impatience with the slow development of residential air conditioning. The Union Electric Light and Power Company in St. Louis, for example, in resolving to stay one step ahead of public demand, drew upon its own experience with the domestic electric refrigerator, which "leaped into public favor so suddenly as to take even the most optimistic by surprise." When a satisfactory self-contained room cooler was not available in 1931, the company set up an experimental laboratory, assembled a prototype, and was poised for production to sell to its customers when the 1932 season brought forward a range of domestic equipment.[88] That season saw the first of the air conditioners—self-contained plug-in appliances with no remote connections.

FROM CUSTOM DESIGN TO PATENT RIGHTS

The machine that established the design for the plug-in air conditioner was developed in 1932 by a Baldwin Southwark subsidiary, the De La Vergne Machine Company. The De La Vergne machine was a self-contained, air-cooled, plug-in room air conditioner that transformed air conditioning into an appliance. It was principally the work of four engineers: Charles R. Neeson, Henry C. Heller, Henry L. Galson, and Hans K. Steinfeld.[89]

A substantial part of their research centered on motors, but perhaps more important, the De La Vergne engineers sought to free the air conditioner from any plumbing connection.[90] By substituting an air-cooled condenser for a water-cooled one, they eliminated the need for a water supply. But they still had to dispose of the excess moisture that precipitated out during the cooling process, without a sewer connection. Neeson solved this by evaporating (or, more precisely, entraining) the condensate in the condenser cooling air and discharging it outside the room during the summer, and evaporating water from a tray and supplying it to the interior during the winter.[91] Strikingly, the De La Vergne design was a heat pump.

Patents show that the De La Vergne engineers worked not only on new components but also on the optimum arrangements for those components. They were after a machine that was, in Galson's words, "quiet," "compact," "relatively inexpensive," "ready for immediate operation," "adapted to have an attractive exterior," and, above all, "economically manufactured in quantity production."[92] It was, from the beginning, a design suited to mass production and mass consumption. The plans for the Baldwin-Southwark machine crystallized around Galson's and Neeson's designs, with some improvements in the detachable connection to the outside, larger volumes of condenser cooling air, and quieter operation.[93] Of lasting importance, this design still formed the basis for air conditioners when the industry resumed production in the postwar era.

Despite this success, however, air conditioners remained peripheral to De La Vergne's parent firm, Baldwin-Southwark. With its primary focus on locomotive construction, Baldwin-Southwark liquidated the De La Vergne Refrigerating Machine Division when orders for engines picked up in 1936. In a move that illustrates the growing importance of patents, the company sold the basic patents to a consortium of air-conditioning companies.[94] As long as custom-designed systems prevailed, engineering expertise was crucial to the production and reproduction of every air-conditioning system. When the air conditioner became a mass-produced appliance, however, engineering design was instead captured in patent rights. Whereas the expertise of designers and engineers had largely predicted company success in the early years, technical design now formed a common foundation for the leading firms, and competitive success came from retail or manufacturing prowess.

THE DECLINE OF THE ATMOSPHERIC GUARANTEE

Just as the technology changed in response to the residential market, so did the sales traditions associated with the custom-design era. In 1934 Fiske opined that "the trend now is towards smaller units and less engineering, and more selling is called into play."[95] Carrier Corporation sought a way to reconcile the new device with older company practices of custom design and engineering guarantees. Company executive Bill Prices, concerned with sales, reiterated that it was essential for the Atmospheric Cabinet to be kept free from the "expensive burden of big-system engineering."[96] That meant a change in sales tactics and guarantees. Prices tried to disentangle the new air-conditioning system from the old ideal of man-made weather: "Why is it *necessary* to discuss, state, or guarantee the *exact* conditions of temperature or relative humidity which will be produced in the summer? You are not selling an intricate central system of air conditioning and you are not charging for such a system . . . Do not become entangled in considerations or arguments about the exact cooling (in degrees) or dehumidifying (in %) which the unit will produce in the room in which it is installed."[97] In August 1931 executives agreed that "when guarantees were not demanded," sales representatives should size up the customer's office or room and match it to one of six average types. They could then estimate the volume of the room and use a table to estimate the number of cabinets required to provide adequate cooling. While this method of estimation was based on a series of approximations, Carrier Corporation reconciled itself to the scheme because "it was felt to be essential that some such simple method be available to permit volume merchandising."[98] Thus the company tried to move beyond its engineering roots into the realm of consumer durables.

As atmospheric guarantees withered, air conditioners were no longer closely matched to the weather, architecture, occupancy, or activities that defined the heat load of a room. The self-contained air conditioner permitted the adoption of comfort cooling independent of building design. Increasingly, air-conditioning engineers recommended the use of multiple small air conditioners even for large buildings, if it would be prohibitively expensive to add pipes and ducts to the structure. Soon the public saw the office building or apartment complex with a small unit for every room. The air conditioner opened new markets during the Depression when the lack of new building starts diminished the opportunities to install central air conditioning inexpensively during construction.

But there were disadvantages to separating buildings and their mechanical services. Because air conditioners came in standard sizes and were installed in rooms of varying construction with a variety of outside exposures located all over the country, their performance was variable. Unlike an air-conditioning system, the self-contained air conditioner mitigated indoor environmental conditions rather than controlling them. Air conditioners could improve the environment, but they did not necessarily have the capacity to maintain ideal conditions. The prevailing direction of development had been to incorporate both control and perfection. The laboratories had defined the human comfort zone, and the air-conditioning companies had guaranteed the machinery's atmospheric performance. The window air conditioner lacked both capacity and precision.

When sales representatives estimated the size of air conditioner appropriate for each room, customers could no longer be certain that the machinery had the capacity to produce ideal comfort conditions at all times. The whole notion of atmospheric control within limits defined by the comfort zone was useless if lack of capacity meant that the air conditioner would be operated at full power all the time. The ideal of man-made weather could not survive the elimination of engineering expertise necessary to match the machinery capacity to the heat load.

THE $5 BILLION IDEA

If in the early 1930s Carrier Corporation and other old-line engineering firms felt crowded by the growing competition from refrigerator makers turned room-cooler manufacturers, it was as nothing compared to the press of companies that vied for the consumer dollar by 1935. The boom in air-conditioning companies was partly a response to GE's optimistic forecasts. GE executives were immensely enthusiastic about the possibilities for air conditioning in the residential market, and in March 1932 the company publicly voiced its conviction that home air conditioning represented a $5 billion market.[99]

The pronouncement gave rise to a gold rush of competition from Depression-hungry refrigeration companies, heating and ventilating firms, electrical manufacturers, and newly organized air-conditioning experts. *Business Week* recalled that the "greatest misfortune . . . of the legitimate members of the industry was the five billion dollar idea which brought scores of depression-distraught manufacturers of other things into the apparently green field of air conditioning, there to muddy themselves and the market."[100] By 1935 eighty-one firms were selling some kind of air-conditioning equipment.[101]

"Right now the scramble for recognition is a free-for-all, but indications are that the race will go to the strong, soundly financed and well established contenders," *Business Week* predicted in June 1935.[102] However, the availability of so many diverse products reduced everyone's profits. Willis Carrier remembered that "into this new market the competitors bought their way through price cutting, hoping for a volume that in the end might bring a profit. In this they were disappointed, but the effect on the price situation and upon established air conditioning manufacturers like ourselves was deplorable."[103] By the end of 1935 Carrier Corporation had racked up $2.4 million in operating losses.[104] *Fortune* estimated that only its sales of Aerofin, a lightweight finned copper tubing, saved the company from bankruptcy.[105]

Part of Carrier Corporation's economic woes were due to the formidable competition that the large manufacturers provided, but the greatest part of the problem in 1935 was that products varied so widely from standard notions of air conditioning that consumers became confused and skeptical. One engineer pointed out that

> today practically everyone is ready to talk about air conditioning as a result of their contacts with it on the trains, and in stores, restaurants and theatres. Most of these people would define air conditioning as summer cooling. They do not know what a real air conditioning system will and should do. This is one of the principal causes of the uncertainty and confusion in the customer's mind when too often they receive prices varying as much as 200 or 300% for equipment that presumably will provide air conditioning.[106]

One analyst noted that "no sooner had the public been informed of the advantages which accrue from air conditioning than a crop of cheap 'air conditioners' appeared on the market consisting of equipment, or lack of equipment, to be bought eagerly by thousands looking for the benefits of conditioned air without the expense."[107] The ideal of man-made weather was crumbling: First went the precise control of humidity and the spray chamber of the air washer, then the atmospheric guarantee and the engineering expertise required to sustain it, and last the public recognition of humidity as an essential element of comfort air conditioning.

A flurry of products for residential cooling appeared on the market, most of which sacrificed humidity control to affordability. One of the strongest of the partial systems was ice cooling. The ice industry was suffering not just from the Depression but also from the growing popularity of the electric refrigerator. Statistics showed that the ice industry continued to increase production until 1932, but it was impossible for observers not to see that ice was being surpassed by mechanical refrigeration.[108] Indeed, by 1936 half of the utility customers in St. Louis owned an electric refrigerator.[109] Refrigeration engineer R. T. Brizzolara conceded that "the ice man no longer plies the streets of the fashionable residential districts, and . . . the feeling goes round that ice is somehow out of date."[110] The loss of older markets put ice companies in the same predicament as companies that had lost their industrial customers as a result of economic conditions. Noting that everyone seemed to be chasing the same market, engineer Samuel Bloom expressed his opinion that "the machinery manufacturers have made such a start in the air conditioning field that it is doubtful whether the ice industry can ever capture any substantial portion of the business."[111]

134

Looking for new customers, ice companies emphasized that their product provided an affordable alternative to the expensive room coolers with refrigeration units. Brizzolara argued that only through the affordability of ice would cooling be brought to the small consumer. "Air cooling has its application at present only where crowds gather, and without a much cheaper system, which ice alone can provide, there is not the least chance of making the market universal."[112]

The cheapness of ice was linked with the short cooling season in the Northeast. The Union Ice Company argued in a 1936 advertisement that air conditioning was "even better business when economically installed—without a heavy year-round investment to meet a short seasonal need."[113] C. P. Yaglou at Harvard's Department of Illumination and Ventilation attached figures to that argument. He estimated that in the northeastern United States the cooling season was twenty to thirty days a year scattered throughout the summer; where ice cost $3–$4 a ton, it was less costly to use ice than mechanical refrigeration.[114]

Research on ice cooling at the University of Illinois test residence, however, showed a more complex situation. In 1933 the Illinois Ice Industries Association sponsored research at the university to test the feasibility of ice cooling. At Test Residence No. 1, occupied by researcher Seichi Konzo and his wife, Kimi, an ice storage system in the basement fed a cooling coil in the furnace duct with melted ice water. While the engineering team gathered technical data on the air cooling, they also consulted with Kimi on the housewife's reaction. The tests showed that it required 2 tons of ice a day to cool the house. Konzo estimated the cost of ice at twice Yaglou's figure, but whether it cost $8 or $16 a day to feed the ice machine in the basement, the figure was still high.[115] Although small electric plants produced some economies in the production of ice, delivery remained expensive.[116] Ice could be produced for under $2.50 per ton, but delivering it across the sidewalk to the customer cost an additional $2.50–$4.50 per ton.[117] The Illinois Ice Industries Association was properly discouraged.

Many of the systems that were marketed in the air-conditioning boom of the 1930s offered considerably less environmental control than ice-cooling systems did. One critic complained hotly that "most of these so-called air conditioners consist of nothing more than a fan built into some sort of fancy cabinet and selling for about three times the value of the materials used."[118] *Architect and Engineer* declared that "until someone makes a reasonably thorough review of what air conditioning is and what is required to produce it, most anything may be offered from exhaust fans to small vaporizers."[119]

REDEFINING AIR CONDITIONING

If the leading makers of residential air conditioners were surprised that consumers failed to distinguish between true air conditioners and competing products, they had only to remember that the industry was reaping the legacy of nearly two decades of theater installations. Generally, both theater management and audiences thought of air conditioning in terms of cooling and not of humidity control. The confluence of refrigerator makers and air-conditioner manufacturers did not help disentangle those notions. When Frigidaire invoked its established reputation to reach new customers, telling them that "the Frigidaire System actually *refrigerates* the air—*just as your household Frigidaire does,* using the same efficient, time-tested type of cooling unit," it only reinforced people's prejudices.[120] One engineering journal understood that the engineers' focus on humidity control and the public's concern with simple cooling would create problems when trying to transform air conditioning into a consumer product: "To the layman who wants to cool off in his own small way it is a help to speed up the air and something better to lower the temperature . . . Whether the relative humidity will be any better is a question, however. The effort being made should start the long job which is going to be necessary before air conditioning in the home comes into its own—that of making the public humidity conscious."[121]

In response to the egregious claims of companies like the one that sold a 6-inch exhaust fan as "air conditioning for the kitchen," eleven companies banded together in 1935 to write a code.[122] These included Carrier Corporation, De La Vergne Machine Company, Frigidaire Corporation, General Electric Company, Kelvinator Corporation, J. H. McCormick and Company, John J. Nesbit, Inc., Parks-Cramer Company, B. F. Sturtevant Company, Westinghouse Electric and Manufacturing Company, and York Ice Machinery Company.[123] Their efforts appear to have guided the Better Business Bureau's announcement the next month of a definition of air conditioning. Air conditioning in the summer, the bureau declared, should cool, dehumidify, and circulate the air, and in winter it should heat, humidify, and circulate the air. Gone was the insistence upon filtering or washing the air, for virtually none of the equipment on the market contained a spray chamber. Yet even by these standards only about half of the eighty-one firms in the business could furnish equipment that would qualify as "all-year round" by Better Business Bureau standards.[124]

Business Week confidently predicted that "this early effort will help substantially in preventing unscrupulous concerns and promoters from gaining a foothold in the industry."[125] The following year the Federal Trade Commission stepped in to regulate the controversy and reprimanded one manufacturer of an air-purifying and air-circulating device that had been advertised as an air conditioner. The manufacturer was ordered to discontinue use of the term *air conditioning* in its advertising. The commission defined true air conditioning as consisting of warming, humidifying, and circulating of the air in the winter, and cooling, dehumidifying, and circulating in the summer, and preferably also cleaning of the air at all times.[126]

The continuing insistence of the largest firms on the inclusion of humidity regulation as an integral part of air conditioning comported with the established definition of the technology. But the switch from the spray chamber to the cooling coil, which characterized all the new room coolers, was accompanied by a change in how air-conditioning manufacturers thought about humidity control. For the most part, air conditioners no longer set humidity levels independent of the temperature. The cooling of the air might reduce the humidity as atmospheric moisture condensed on the cooling coil, in the same way that droplets of moisture form on the outside of a cold glass of iced tea, but seldom was provision made to regulate the humidity level of the air as it left the air conditioner, and never could the company predict the humidity levels that would occur in the conditioned room. Nearly all room and window air conditioners failed to provide the same environmental control that central-station air conditioning could produce. They lacked the capacity and the sophistication to regulate humidity levels. They might provide more humidity in the winter or less humidity in the summer, but the machines generally did not regulate humidity levels to a set standard. The insistence on humidification and dehumidification at this point had a great deal more to do with disciplining competition in the marketplace than with maintaining conditions within the comfort zone or sustaining the ideal of man-made weather. The drive to include humidity in the definition of comfort air conditioning in the theaters during the 1920s was vastly different from the inclusion of humidity in air conditioners in the 1930s.

THE IMPORTANCE OF MASS PRODUCTION

The promoters of air conditioning hoped the combination of a new technology and the American mass-production system would lower prices, create jobs, and build profits. Refrigeration engineer Everett Ryan offered his prescription that air conditioning "must build up standardization and mass production to a point approaching that which we now have in the case of the electric refrigerator."[127] Mass production and marketing was a natural move for the large refrigerator makers like Frigidaire. In 1935 Frigidaire abandoned its previous method of unit construction and put two endless-belt production lines into operation.[128] York Ice Machinery Corporation also put air conditioning on a regular production line about the same time.[129]

Carrier Corporation moved increasingly into manufacture as well.[130] In 1937 the company acquired a factory in Syracuse, New York, that had been seized for $500,000 in unpaid taxes. Encouraged to relocate by city officials and businessmen, Carrier Corporation paid a token $1,000 at auction for the $4 million factory. It consolidated its five factories at the new location and trebled its floor space besides.[131] By 1938 Carrier Corporation's income was split evenly between manufacturing and custom installations.[132]

Few of the new air-conditioning companies made all their own components. Only about ten old-timers made their own equipment. For a few companies, manufacturing for the trade was an important activity. Trane Company of La Crosse, Wisconsin, for example, made virtually every type of air-conditioning equipment, particularly cooling coils, which it supplied to other air-conditioning companies, including GE.[133] A review of the field showed that 84 percent of the firms that sold air conditioning bought pumps from external sources for their equipment, 78 percent bought motors, 64 percent bought compressors, 55 percent bought cabinets, 41 percent bought air washers and filters, 41 percent bought dehumidifiers, 41 percent bought heaters, 38 percent bought fans, and 24 percent bought humidifiers.[134]

Walter Fleisher found little to recommend the new mass-produced equipment. He felt that standardization of air-conditioning equipment could not respond adequately to the multiplicity of climatic conditions across the United States. In 1935, in a talk before the Engineering Societies of Canada, he put forth the view that "modern quantity production machinery was designed for the densest market . . . and that then, in an effort to dispose of the surplus material, every other section of the country was pestered and pounded to consume the same equipment—whether it fitted their particular requirements or not. Air-conditioning equipment, particularly of the unit type, has been developed for the New York and Chicago areas, but is being pushed in other localities, irrespective of their needs."[135] If the window air conditioner severed the connection between the machinery and its building, the new style of production further separated consumption from geographical need.

A PRIME PUBLIC DISAPPOINTMENT

By the end of the decade, critics were reassessing the potential of comfort air conditioning. The expectations had been unattainably high. *Fortune* remembered that "the public, led on by the exaggerated pipe dreams of Sunday supplement writers, looked for a huge and immediate mass market . . . it looked, in the not too far future, toward a revolutionized indoor life on this planet in the manner of an H. G. Wells fantasy." Instead, the magazine called air conditioning "a prime public disappointment of the 1930s."[136] *Time* noted sadly in 1937 that "touted as one of the new applied sciences that were going to bust the slump, air conditioning is still small potatoes."[137] Indeed, with 22 million wired homes in the United States in 1938, less than 0.25 percent had air conditioning in one or more rooms.[138]

A magazine survey asked consumers if they would air-condition their house if it cost $1,200 or one room if it cost $200. These target figures were lower than prevailing prices; air conditioning for an entire house cost approximately $1,500, and a room cooler cost $400. Yet survey respondents were unenthusiastic; two-thirds said they would not choose to air-condition, a proportion that did not vary across income groups. "So long as it costs $400 or more to buy a portable cooler for one room, most people are going to trade in their old cars for new ones instead," *Fortune* predicted.[139]

But the surprising finding that income made little difference in respondents' willingness to buy air conditioning suggested a more general problem with public recognition and acceptance. "Most people think it is a fancy name for making rooms cooler in the summer," a popular magazine reported.[140] Another conceded that "while the term includes humidity control, air movement, and heating as well as cooling, popular usage puts the emphasis on cooling."[141] "With the major emphasis placed on cooling in railroad, theater, and store," *Business Week* reminded readers, "controls were far too often set for temperatures under the human 'comfort zone.'"[142]

Despite the enthusiasm for comfort air conditioning during the Depression years, it seems that many residential installations of the 1930s failed to impress consumers. Housewife Alice Thalman recorded just such disappointment with her newly air-conditioned house. On a day when the outside temperature reached 81 degrees, Thalman seated her bridge party guests in the air-conditioned living room, where the temperature was 80 degrees but the humidity was low. Unfortunately, her guests did not sense any relief from the lowered humidity and soon suggested a move to the terrace off the sun porch.

Disappointed in their reaction, Thalman believed that her guests had failed "to recognize immediately their relief from the oppressive conditions out of doors" because "we had all been in the habit of patronizing a certain theatre in town wherein the air was cooled not conditioned."[143]

The theater-educated public repeatedly failed to acknowledge or care that humidity control was an important feature of comfort cooling. This indifference opened the door to a variety of new designs that had in common their divergence from the engineering community's vision of man-made weather, with all its emphasis on control and perfection. The centrality of humidity control, and the necessity of precise humidity control to human comfort, faded with the spray chamber and the bypass circulation system. The window air conditioner reflected not the engineering rationality of matching the machine to the building, climate, and people, but the consumer values of comfort, mobility, gadgetry, and ownership. This transformation broke the exclusive hold of engineering firms and their ability to dictate the functions and form of the technology.

CHAPTER SIX

From a Luxury
to a Necessity, 1942–1960

The custom-designed, central air-conditioning system represented one extreme of the spectrum of design possibilities, and the mass-produced window air conditioner the other. Air-conditioning firms that were engineering-oriented favored high-performance machines and close environmental control, while electrical appliance manufacturers promoted affordability and a pragmatic attempt simply to take the edge off the heat. Throughout the 1930s these competing visions of the still evolving technology pulled consumers in opposite directions. The shape of future development was not yet settled by the end of the decade, and when the attack on Pearl Harbor pitchforked the United States into World War II, the debate was set aside.

In fact, those in charge of organizing American production for war pushed comfort air conditioning out of the factories and denied it scarce materials. The War Production Board issued Order L-38 in May 1942 prohibiting the installation of new systems or the manufacture of new equipment solely for personal comfort.[1] Process air conditioning, however, was recognized as an important aid to military production. Thus, air-conditioning manufacturers continued to build limited amounts of industrial equipment for the war effort at the same time as they turned out machine guns, airplane wings, and antisubmarine equipment.[2]

To express the nation's new priorities in the face of wartime needs, highly publicized plans were made to remove existing comfort systems in civilian and government buildings and install them in factories engaged in military production. Indeed, a few actually were moved. In New York City a centrifugal refrigerating machine was taken out of Tiffany's and installed in a B. F. Goodrich plant in Port Neches, Texas, and the equipment from Lord and Taylor's was shipped to Sinclair Rubber in Houston. Stores in Washington, Chicago, and St Louis pursued a similar policy.[3] Such sacrifice must have been good publicity for these highly visible retailers. In contrast, Washington legislators were extremely reluctant to give up the numerous comfort plants that had been installed in Washington buildings in the preceding decade.[4] Little was said publicly about the relative contribution of bureaucrats to the war effort, yet actions often speak louder than words, and bills that authorized the transfer of government-owned equipment to war plants invariably died in committee. Finally, the entire program to redistribute equipment was abandoned in 1944.[5]

In this new ascetic atmosphere, it even proved difficult to keep existing plants in operation. Refrigerants for recharging such systems were in very scarce supply, and equipment that used Freon-12 were particularly hard hit. The addition of new manufacturing facilities for Freon in the spring and fall of 1944 quadrupled the peacetime productive capacity, but output was still limited by the shortage of hydrofluoric acid, a necessary component.[6] In this world of constraint, users scrambled for the limited amount of Freon available, especially since large quantities of the chemical had been allotted to the army's antimalarial campaign for use as a propellant for insecticides.[7] Given the jealousy with which the federal bureaucracy guarded its own comfort systems, owners of enfeebled air-conditioning systems grumbled that their applications for a rejuvenating charge of Freon were probably denied by some bureaucrat occupying a "swanky, air-conditioned Washington office." But Loring Overman, *Heating and Ventilating* magazine's Washington reporter, wrote that the air-conditioning division of the War Production Board was housed in a temporary stucco building left over from World War I, and that the "sweltering members of the A-C Section staff signed each denial in perspiration as well as ink."[8]

While theaters, bowling alleys, and other customers waited for the release of small amounts of refrigerants to keep their comfort systems operable, the industry dreamed of the postwar market. Manufacturers hoping for a peacetime mass market worried about the effect of air conditioning's image as a luxury item. The characterization of comfort air conditioning as an indulgence had been central to its appeal in the 1910s and 1920s, when it was the ultimate luxury offered by opulent movie theatres. That perception had been reinforced during the war years, but with a negative spin, when the rhetoric of personal sacrifice reminded Americans that individual lifestyles were directly connected to the nation's war effort. So the industry fretted that postwar buyers might view air conditioning not simply as a temporarily expendable frill but as an unjustified extravagance. Industry representatives wondered whether the connotation of luxury would boost or inhibit sales.

Many industry analysts declared consumer attitudes to be the greatest barrier to adoption of comfort air conditioning. Accordingly, one industry vice president outlined the difficulty to a sympathetic audience: "While we have been learning since the 1930's, to expect it in public buildings and to want it in our homes, many of us still consider air conditioning as less than essential or at least less than necessary."[9] Another advocate concurred: "The problem is to convince the buying public that cooling is just as logical an addition to their dwelling place as is heating."[10] Yet a third voice expressed the same sentiment: "The time seems not far distant when mass production will bring air conditioning to every room of low cost houses. The problem has been one of selling the public on the idea that air conditioning is no longer a luxury."[11]

Surprisingly, manufacturers discovered in the postwar era that consumer approval accounted for only a small part of air conditioning's postwar success. Instead, acceptance of the new technology by all segments of the construction industry proved to be a vitally important factor. A blend of new architectural designs and new commercial building practices placed air conditioning at the heart of postwar homes and office buildings. Architects and builders made the decision to air-condition American homes and offices, then lenders and regulators stamped the change with institutional approval. In these years air conditioning was reconceptualized from a luxury to a necessity—but not simply because mass production, cheaper prices, and consumer enthusiasm made it so. Contrary to the expectations of the 1930s, when analysts predicted a replay of that American miracle whereby Henry Ford had transformed the automobile industry, the triumph of air conditioning in the 1950s was based upon its integration into building design, construction, and financing. Architects, builders, and bankers accepted air conditioning first, and consumers were faced with a fait accompli that they had merely to ratify.

PICKING UP THE PIECES

At first, there was little recognition that the postwar years might be distinctly different from what had gone before. Air-conditioning manufacturers appeared to pick up the pieces from 1942—the same technology, the same assumptions, the same animosities—and carry on.

Following victory in Europe, the limitation and conservation orders of the War Production Board were rapidly repealed.[12] In particular, the revocation of Order L-38, governing the air-conditioning industry, allowed manufacturers to resume production for the first time since 1942.[13] And because it was where they had left off, as soon as materials and manpower were available, the industry responded with a reissue of the 1942 console-style room air conditioners.[14]

In the postwar era, hopes for the industry mirrored the aspirations of the 1930s. Many continued to believe that mass production would make air conditioning finally affordable. *Life* prophesied in 1945 that "one of the prewar luxuries which seems most likely to come out of the luxury class and into the postwar mass market is air conditioning."[15] The magazine repeated the conviction that mass-production methods would bring costs within the reach of the average consumer. By now, however, this ambition had a patina of age; Willis Carrier thought the moment had come in May 1929, before the stock market crash, when he predicted that "in years to come, air conditioning and cooling for summer may become a necessity rather than a luxury, and we will look upon the present times as marking the end of that 'dark age' in which there was but relatively little cooling for human comfort."[16] The failure of air conditioning to reach its expected commercial maturity in the intervening fifteen years had done nothing to diminish supporters' faith in its potential.

Industry ambitions were seemly fulfilled by a steady climb in sales. In 1945 just over 1,000 room air conditioners were shipped; the next year, nearly 30,000.[17] By 1950 production had increased sixfold to 193,000, and it climbed to 1.3 million by 1956.[18] Although air-conditioning production increased rapidly after the war, it began its rise from nearly zero. Thus, each year manufacturers achieved remarkable percentage increases, yet a decade elapsed before the volume of sales began to match the extravagant rhetoric about air conditioning's potential.

As predicted, prices did decline during the 1950s, and consumers reaped the benefits of quantity production. The smallest portable room conditioners cost nearly $400 in 1938; by 1952 they ranged in price from $229.95 for the smallest (⅓ horsepower) size to $419.95 for a ¾-hp model.[19] Although the market was fragmented among more than 100 brands in 1953, economies of scale were achieved in part because of the concentration of production. Only about a dozen firms actually manufactured air-conditioning equipment. Large manufacturers, such as the Hotpoint Company, Philco Corporation, and RCA, were among the crowd of competitors for the home market, but these firms bought their equipment from O. A. Sutton Corporation, York Corporation, and Fedders-Quigan Corporation. Fedders-Quigan alone made roughly one-quarter of all the air conditioners sold.[20]

Indeed, Fedders-Quigan's president, Salvatore Giordano, "put all his eggs in the room unit basket."[21] In doing so, he seemed to have the weight of historical precedent behind him, which demonstrated that mass production, cheaper prices, and consumer demand were the path to success and profits. It was a business strategy that had captured the imagination of hundreds of firms during the 1930s, and many returned to it in the immediate postwar era.

That commitment to self-contained room conditioners put the company in conflict with Carrier Corporation, which still favored central air-conditioning systems for the home.[22] A pioneer of custom-designed systems that had the power and sophistication to produce artificial indoor climate, Carrier Corporation still believed that homes could and should be equipped with central air-conditioning systems. A joint program with Lennox during 1943–44 had confirmed Carrier executive W. R. Hill in the belief that "relief cooling" was insufficient and that "a full scale comfort job is the only thing that will satisfy the customer."[23] One observer noted that the corporation acted as if "the room air conditioner is merely a necessity during a period of transition in the development of the air conditioning market."[24]

Carrier Corporation believed that central air conditioning could be economically installed in American homes to maintain control of humidity and temperature within the comfort zone. Beginning in 1950 the firm began to size residential equipment differently to make central air conditioning less costly. Studies revealed that the temperature in an unheated and uncooled house generally fluctuated by only 7 degrees, even when the outside temperature ranged as much as 33 degrees. Accepting a 7-degree variation as normal, company engineers settled on 74 degrees as ideal but allowed the early morning temperature to dip to 72 degrees and the evening readings to hit as high as 78 degrees. Called 24-hour estimating, this acceptance of a 7-degree fluctuation over 24 hours enabled Carrier Corporation to decrease the size of the machinery, perhaps by as much as 50 percent, and reduce the initial cost to the consumer accordingly. Under the new system, the machinery operated continuously and the temperature varied over the course of a day, but presumably remained within the limits of the comfort zone.[25] And, Carrier Corporation hoped, the new system would also fall within the limits of the consumer's budget. Company officials estimated that a customer would pay $2,050 for a 3-ton air-conditioning system and that the cost of operating it for a cooling season would run from a high of $124 in Dallas to a more modest $53 in St. Louis.[26] By

1955 Carrier Corporation had used 24-hour estimating for 5 years in 10,000 in-stallations.[27] That year the company urged its dealers to sell the idea of build-ing air conditioning into the house rather than installing it in the window.[28]

Thus the fissure, opened in the 1930s, widened in the 1950s as companies de-scended from engineering firms delineated their interests separately from those that began as electrical manufacturers. The former continued to pro-mote a vision of air conditioning as an extended technological system that was mated to the building it served and tied to it with ductwork; the latter persisted in extolling the convenience of air conditioning as a plug-in appliance that easily followed its owners from one residence to another. *Business Week* re-ported in 1957 that "manufacturers are still divided on where the big home market lies." The magazine acknowledged that "the central system camp still is convinced that central systems represent the greenest pasture in the long run. But there's a new spirit of optimism among room unit makers."[29]

The most visible manifestation of this revived conflict over technical design came a year later. In May 1958 GE led Fedders-Quigan and Hotpoint in a resig-nation from the room air-conditioner section of the trade association, the Air-conditioning and Refrigeration Institute. The air conditioner was an appli-ance, these companies maintained, and should be merchandised as such. Room air-conditioner manufacturers resented the implication by the rest of the in-dustry that central-station systems afforded better air conditioning than room conditioners. They saw it as being in their best interest to establish a room air-conditioner section in the National Electrical Manufacturers Association instead.[30]

The debate over central air conditioning versus room conditioners showed that Carrier Corporation was no longer in a position to dictate the accepted form of air-conditioning systems. Engineering expertise was not essential to the mass market, and Carrier's patent position in the comfort air-conditioning industry had changed dramatically in 1945 with the dissolution of the Audito-rium Conditioning Corporation. From the founding of Auditorium Condi-tioning, the centrality of the bypass patent to any economical large-scale in-stallation had ensured its virtual monopoly of the comfort-conditioning field and had given rise to much grumbling. In 1931, for example, theater executive Barney Balaban had complained bitterly that "he did not feel it right to give in to something which amounted to the same thing as a monopoly which in his opinion was the case insofar as the bypass was concerned."[31]

Consulting engineer Charles Leopold, too, had resented Auditorium's stranglehold on economical comfort systems. In 1931 he had carefully designed the air conditioning for the Saks and Gimbel's department stores in New York City so as to avoid the $13,000 bypass royalty fees.[32] Leopold, who designed sixty air-conditioning installations for Warner Brothers Theatres, was one of the most adamant and successful opponents of Auditorium's claims. When Auditorium filed a patent infringement claim against Warner Brothers Pictures and Warner Brothers Theatres in July 1931, Leopold helped with Warner Brothers' successful defense that the Klein patent (No. 1,296,968) anticipated the bypass and invalidated Auditorium's claims.[33] The U.S. District Court agreed with Warner Brothers and dismissed Auditorium's complaint on 12 June 1935. Auditorium appealed in January 1936, but the decree was affirmed in April.[34] Despite the loss of its suit, Auditorium threatened legal action against Warner Brothers, Loew's, and National Theatres Corporation (Twentieth Century–Fox), and won an out-of-court agreement from those companies that the bypass royalties would continue to be paid.[35]

Thus, Auditorium continued its dominance, much to the benefit of its constituent members, Carrier Corporation, B. F. Sturtevant, York Ice Machinery Corporation, Ross Industries Corporation, and American Blower Corporation.[36] The company's success made it quite visible, however, and in August 1943 the Anti-Trust Division of the Justice Department filed against the company for effecting a monopoly in restraint of trade in the manufacture, sale, and installation of air-conditioning equipment. Government critics contended that Auditorium received royalties on 90 percent of air-conditioning installations. In 1945, after nearly twenty years of business, the company was ordered dissolved and the patents dedicated to the public. Lawrence S. Aspey, chief of the New York office of the Anti-Trust Division, declared that "in the public interest an effective monopoly has been . . . dissolved. Competition and mass production should for the first time bring air-conditioning equipment within the reach of everyone's pocketbook. The small homes and factories of the nation will soon enjoy benefits which are the direct result of our three-year tussle with the air-conditioning industry."[37]

No longer in a commanding position with regard to engineering expertise or patents, Carrier Corporation faced stiff competition. The fracture of business power among a large number of competing firms brought a proliferation of air-conditioning designs. Or, just as important, technological innovation had opened the door to many new firms in the 1930s, and they consolidated their position in the 1950s. The growing market strength of room air conditioners and their manufacturers provided an important challenge to the ideal of man-made weather that Carrier Corporation and other pioneering companies had promoted and developed for decades.

Indeed, room air conditioners became quite popular and far outstripped central air-conditioning systems in number. In Chicago, for example, Commonwealth Edison Company reported that 57,000 room air conditioners were added in 1955 as opposed to an estimated 1,000 central systems.[38] The vast majority of room conditioners were purchased by individuals for home or apartment use and were reportedly installed in bedrooms for more comfortable sleeping.[39] Room air conditioners were a logical choice for consumers who lived in rented quarters or in older housing stock that was expensive to retrofit with central air conditioning.

Central air-conditioning systems were identified, instead, with new construction. With the need for ducts, plumbing, sewer connection, and adequate electrical wiring, central systems were far less costly when built into a house at the time of construction. Retrofitting an existing house could be prohibitively expensive. The small number of housing starts during the Depression had provided limited opportunities for these systems to be installed affordably, although a few Depression-era housing developments did include air conditioning as a standard feature. Unusual at the time, these developments presaged several solutions to the vexing problem of how to make central air-conditioning systems affordable for the mass market.

In 1935, for example, Chrysler Airtemps had joined Friendship Homes to build eighty fully air-conditioned homes on Long Island.[40] The standardization of house design allowed builders to standardize engineering design as well—using essentially the same heat load, air-conditioning equipment, and duct layout for eighty homes. Developers were still nervous about the additional cost of central air conditioning over a non–air-conditioned home. In an attempt to balance the advantages and liabilities of built-in air conditioning, architect-builder Waverly Taylor in 1935 adapted designs from General Electric's "New American" homes competition to build seventy-three houses, twenty of which included the ductwork and plumbing for a central air-conditioning system, but not the expensive compressor.[41] Thus, standardized engineering design and "roughing in" the air conditioning in the construction phase provided two solutions to the problem of the high cost of central systems.

It soon became clear that postwar conditions would favor the expansion of central air-conditioning systems. The expected rush of new construction for returning servicemen and their families provided a golden opportunity to incorporate air conditioning into American houses from the start, for a fraction of the cost necessary to retrofit older homes. Those in the air-conditioning industry who favored central air-conditioning systems felt that their future was "inalterably linked with the conventional building industry."[42] They must have realized that connecting their product to new home construction was the most effective way to match the lowered costs that window units achieved through mass production.

AMERICA'S POSTWAR BUILDING BOOM

Thus the success of central residential air conditioning was linked to a rise in new home starts. Although the need for more housing was clearly recognized by all, the exact solution to the housing problem drew no such unanimous agreement. After the tight regulation of the war years, the construction industry was determined to return to business as usual, unfettered by government restrictions. The federal government was equally resolved to compel the marketplace to favor returning veterans and their families.

The provision of veterans' housing quickly became a federal objective. Although the government dramatically reduced its regulation of private construction with the revocation of Order L-41 in October 1945, it was not inclined to let market forces of supply and demand shape the housing available to returning veterans. In December 1945 the national housing administrator, Wilson W. Wyatt, set ambitious goals for the amount of residential housing to be constructed in 1946 and 1947.[43] He targeted the completion of 1.2 million new residential units in 1946 and 1.5 million in 1947—goals far above the existing record of 937,000 units of new residential construction, set in 1925.[44] To reach these new heights, the government proposed to convert wartime buildings, sponsor temporary and prefabricated buildings, and most important, channel scarce building materials into low-cost veterans' housing. At first, veterans' housing was allotted 35 percent of available building material, but in April that was increased to 50 percent, a much larger percentage than residential construction had ever before absorbed.[45] The Veterans' Emergency Housing Program, begun early in 1946, focused on inducing private industry to build low-cost veterans' housing (under $10,000).[46] An inexpensive house generally meant a small house, and indeed the Federal Housing Administration (FHA) circulated six floor plans for small veterans' houses, none of which exceeded 653 square feet.[47]

In the face of growing opposition to Wyatt and to the office of national housing administrator, the federal government exercised its influence principally through loan policies rather than direct intervention. Long after the removal of restrictions on building materials, the federal government's commitment to long-term, government-insured mortgages encouraged the construction of low- and moderate-cost housing. By mid-1947 one out of every twenty veterans had taken out a G. I. loan.[48] With an average annual income of $3,619, veterans bought houses with an average valuation of $6,545.[49] Such modest homes were not the typical products of most builders; a survey revealed that while some were selling houses in the $5,600 range, the average sale price hovered around $9,000.[50]

Private industry explored two distinct approaches to affordable housing. One was squarely within the political economy of the commissioned house, in which homeowner and architect in partnership built a house suitable for the owner's needs, budget, and site. Within that traditional framework, some architects in the postwar years argued that savings could be achieved through new types of designs. The second approach put building on a different economic footing altogether. Tract developers tried to lower costs through new construction methods made possible by large-scale projects. In the end, the typical postwar house incorporated both new design and new construction methods.

Architects, of course, emphasized the importance of design. They had always argued that the professional expertise of the designer saved the owner the costs of bad design decisions, but that general justification now gave way to the specific claim that an architect was best trained to avoid the pitfalls of government restrictions and to take advantage of mass-produced materials. In response to the nation's housing crisis, architects argued that intelligent design was more important than ever in low-cost homes. To illustrate that point the Los Angeles–based magazine *Arts & Architecture* commissioned a series of designs for the small house beginning in 1945. The magazine explained that the project was "an attempt to find out on the most practical basis known to us, the facts (and we hope figures) which will be available to the general public when it is once more possible to build houses."[51] This effort became known as the Case Study House project.[52] The designs were all for modest, servantless, single-family houses, and the first, designed by J. R. Davidson and built in 1946 under government restrictions, encompassed only 1,100 square feet.[53] They gained a wide exposure due to the magazine's decision to build the Case Study designs as model homes. Half a million people visited the first dozen houses that opened for public viewing. *Architectural Review* called the effort "one of the most distinguished and influential architectural research programs ever inaugurated."[54]

The project tapped the talents of several notable architects, and the two houses designed by Charles Eames and Eero Saarinen were distinguished by the concerted effort to use industrial materials both for economy and for a modern aesthetic. The Case Study Houses captured, early in the postwar years, many of the elements that distinguished residential design for the next two decades. The project "demonstrated that a good house can be of cheap materials; outdoor spaces are as much a part of the design as enclosed space; a dining room is less necessary than two baths and large glass areas; a house should be turned away from the street toward a private garden at the back."[55]

Such features came to characterize almost all small postwar houses, whether built by the homeowner or the tract developer. Most important for the air-conditioning industry was the relationship between the house and the outdoors. The opening of the small, inexpensive house to the outdoors, both visually and literally, had the virtue of creating a sense of space in the altogether too small postwar house. Architects designed with fewer interior walls, and they annexed outside space as an adjunct to the cramped interior—visually, through the use of large picture windows, and physically, through the use of sliding glass doors. The attempt to minimize the boundaries between indoors and outdoors through the removal of visual barriers entailed the use of large expanses of glass. Such a style inevitably produced a house vulnerable to the extremes of hot and cold weather.

The Case Study Houses were aimed primarily at individual homeowners who worked with a contractor to build their own home. Yet this traditional, one-at-a-time approach to home construction seemed poorly suited to meet the pent-up demand from the four-year hiatus on home construction. Government officials like Wyatt pinned their hopes on prefabricated housing that promised to bring the efficiency and savings of industrial mass production to the craft-based construction industry.

It seemed for a time that wartime production genius Henry Kaiser would be the one to apply the factory-production formula to veterans' housing. However, Kaiser decided to follow a more established path. At the end of 1945 he formed a partnership with Fritz B. Burns, a prominent land developer, to establish a $5 million company, Kaiser Community Homes, for the explicit purpose of building low-cost housing. Although Kaiser Community Homes incorporated a 5-acre plant in Los Angeles to fabricate wall, ceiling, and floor panels,

Kaiser and Burns channeled those materials into their own developments rather than using them to build prefabricated houses for the open market. By September 1946 the company announced it was completing 100 houses a week—two- and three-bedroom houses priced from $6,950 to $8,650.[56]

Just as it had to Kaiser, the development of raw land proved irresistible to many postwar builders, and developers, like homeowners, benefited from new loan policies. Central to these new suburban housing tracts was the availability of largely agricultural land just beyond the city center. This raw acreage was bulldozed clean of any existing trees and shrubs to present a *tabula rasa* for the necessary streets, sewers, and sidewalks. In California the Bank of America offered a new type of loan that covered a substantial part of the purchase price of the raw acreage and site improvement costs, in contrast to the traditional construction loan that advanced money only on structural improvements.[57] It was widely publicized by the National Housing Agency.

Indeed, the majority of new houses were constructed by speculative builders who achieved savings by building standardized houses in vast suburban tracts. The best known developer was Abraham Levitt who, with sons Bill and Alfred, used economies of scale, repetitive designs, small power tools, cheaper materials, precut lumber, and new construction techniques to bring the savings of mass production to home construction short of prefabrication. The Levitts, who had built 2,500 wartime housing units in Norfolk, Virginia, easily completed 1,000 veterans' houses in 1946.[58] They were typical of many postwar developers who had learned their skills erecting wartime housing. *American Builder* saw the connection clearly:

> The expansion of nonfarm residential building in the period ahead will be facilitated by experience gained in the war housing program, in which both builders and a few general contractors built projects of several hundred units. With few exceptions, these projects were marked by more thorough planning of operations, more careful timing, and control of materials, and greater use of power-operated tools than were general in pre-war operative building. Several larger contractors accustomed to management procedures were introduced to the field of residential work, and many residential builders learned the possibilities present in large-scale operations. On some of the largest projects, materials were brought directly from manufacturers. In these respects, and others, the house building industry matured substantially.[59]

Speculative builders reduced house prices through a variety of means, but the most lasting legacy for the air-conditioning industry was the emphasis on simple and lightweight construction. The partnership of Ian Murray and Cedric Sanders in southern California illustrates builders' efforts to reduce costs by continuing to incorporate elements that had served them well in wartime construction. Murray and Sanders used concrete-slab floors, drywall construction, low-pitched roofs, exposed cathedral-type ceilings, and a modified post-and-beam type of construction.[60] These elements were popular with many builders.[61] Low-pitched roofs, cathedral ceilings, and lightweight walls (especially those without insulation) provided little barrier to the absorption of heat and could lead to affordable but uncomfortable houses. Combined with a near universal enthusiasm for picture windows, sliding glass doors, and plenty of windows, the postwar house in general looked appealing, cost little, and turned in a very poor environmental performance.

THE AIR-CONDITIONED HOUSE

The postwar building boom created the perfect opportunity for the growth of central air-conditioning systems, but the concentration in low-cost housing was something of an impediment. Even when air conditioning was installed during the construction stage, it still absorbed a large percentage of building costs. In 1946 *Fortune* warned that "among the proposed blessings of the new age of living is complete home air conditioning . . . This is a very flossy blessing indeed, for it is estimated to cost up to $2,000 in a new $10,000 house."[62]

Anxious to incorporate air conditioning in new residential construction, manufacturers promoted the idea that consumers would save money not only from lower installation costs but also from a thorough redesign of the traditional house. If form followed function, then the inclusion of air conditioning should inform every design decision. With an eye to economy, a clever redesign of the traditional house would save enough money to cover the purchase price of the equipment. *Better Homes and Gardens* expressed this view succinctly: "though year-around air conditioning costs more than a system that provides only heat, planning your house for air conditioning will make up much of the difference."[63]

Such a house appeared in *Better Homes and Gardens* in the Five Star Homes series, a set of designs commissioned by the magazine to give its readers "maximum comfort, convenience, and beauty—at minimum building cost."[64] For $5,000, the magazine offered readers the architect's design, detailed working drawings, materials list, complete specifications, and contractor-owner agreements. At least sixty-six designs were produced and meant to have wide circulation. When Servel commissioned an air-conditioned house for the Five Star Homes series, architect David S. Barrows reduced the cost of the installation by laying out the rooms to use a minimum of expensive ductwork.[65] Barrows's air-conditioned house was different in other ways as well, most notably in its independence from window ventilation. Windows still were useful for letting in light, but they no longer needed to open. *Better Homes and Gardens* informed readers that "in houses planned for air conditioning, you can omit basement, screens, and moveable sash to offset extra cost of equipment."[66]

Few proponents of air conditioning advocated placing the equipment into conventional houses; almost all suggested architectural changes.[67] As early as 1946 the air-conditioning industry made an effort to convince speculative builders that air conditioning would not only attract buyers but also would pay for itself by permitting the elimination of several expensive features of traditional design.[68] One magazine reported that "this brisk talk is directed by the air conditioning industry less to the ultimate consumer than to the speculative builder. The latter is assured that, over the long haul, air conditioning will help him sell more houses at better prices . . . the builder may as well learn now to make the whole house absorb the cost."[69] Builders could save money by eliminating movable sashes and screens, and the ell, a wing at right angles to any room that promoted cross-ventilation. These design elements that provided cooling and ventilation were no longer needed now that those functions were to be mechanized. This attempt to economize by redesigning the house around air conditioning depended on the substitution of a mechanical system for architectural features that provided passive cooling and ventilation. Such new houses required air conditioning to make them comfortable.

Air conditioning began to appear in tract houses in 1952. Near Falls Church, Virginia, the Valley Brook development of 80 homes designed by architect David S. Oman and built by Oman-Neal, Inc., included air conditioning in homes priced under $30,000.[70] In New York, a Long Island development called Birchwood at Westbury contained 450 air-conditioned homes.[71] In West Orange, New Jersey, a builder of 89 low-cost homes kept air conditioning affordable by using a small (1 hp) air conditioner that could be directed either to the living room or the bedrooms.[72] Farther south, even less expensive homes experimented with air conditioning as a standard feature: A 210-unit development in Dallas, designed by architect George N. Marble, included air conditioning in its six-room house priced at $12,500.[73] This was followed by two more Dallas developments, a 125-house venture and a 76-house tract, both equipped with year-round air conditioning.[74]

At least two developers in 1952 reported adopting the air-conditioning industry's suggestion of modifying traditional designs to provide air-conditioned models at the same cost as a conventional home. Charles A. New-bergh, president of Heathcote Heights Building Corporation in New Rochelle, New York, planned 75 air-conditioned homes. Heathcote saved money by eliminating attic fans, screens, storm windows, and cross-ventilation.[75] In addition, London Homes of New Orleans built the pilot home of a 500-house project. Designed by Rene F. Gelpi, the homes were priced between $18,250 and $19,300. London Homes estimated that the elimination of some movable sashes and screens, and abandonment of the attic fan with its consequent steeply pitched roof and hall area, paid for two thirds of the added cost of the air-conditioning system. Savings from the "mass production techniques" of tract development made up the rest. "The combined result is that this home could be sold completely air conditioned for the same price as a conventional home," *Heating and Ventilating* calculated.[76]

Fortune reported in 1953 that "speculative builders have discovered the virtues of air conditioning: it makes for construction savings by eliminating the need for cross ventilation, attic fans, extra corners, and a full complement of screens."[77] Some builders grumbled that "costs aren't quite so thoroughly 'absorbed' as salesmen say," but certainly a redesign of traditional features could pay some, if not all, of the cost of air conditioning.[78] A survey by the National Association of Home Builders (NAHB) reported that 40 percent of the 255 leading builders were planning to market air-conditioned houses in 1953 priced at $15,000 or less.[79]

THE RATIONAL AIR-CONDITIONED HOUSE

The drive to find economical ways to incorporate central air conditioning into new home construction, even in the more modest price brackets, took several forms. Reducing the initial costs to the consumer, whether a homeowner or a speculative builder, by "making the whole house absorb the cost," received the most attention. But most air-conditioning experts had a more comprehensive vision: the redesign of the home around the new technology. Air-conditioning manufacturers and engineers promoted means to reduce both the initial cost and the operating expenses of residential systems. According to industry calculations, the development of designs for an air-conditioned house that was inexpensive to buy and economical to operate was the key to widespread acceptance.

To that end, in the summer of 1953 Ned A. Cole, chairman of the NAHB Committee on Air-Conditioning, approached Werner W. Dornberger, chairman of the Department of Architectural Engineering at the University of

Texas. He proposed research on residential air conditioning that focused on a practical test of equipment and houses under the conditions of everyday occupation. The department responded positively to this overture, and out of the proposal came the Austin Air Conditioned Village.[80] Established in Austin, Texas, in 1954, the village consisted of twenty-one houses built by eighteen different builders, equipped by twenty-one different manufacturers, and occupied by real families.[81]

Repudiating window air conditioners as "one of the saddest chapters in the sordid history of air-conditioning for dwellings," and reporting that "the want-ad columns of most Texas papers are usually full of advertisements endeavoring to sell used window air conditioners which have proven unsatisfactory to their users," the researchers indicated that they were firmly committed to the central air-conditioning system.[82] Their objectives included demonstrating the feasibility of air conditioning in low-cost homes and improving both air-conditioning units and residential design for greater economy. In simple terms, they were concerned to make air conditioning affordable and practical by reducing both the initial and operating costs.

The Austin Air Conditioned Village was not the first appearance of the drive for what might be called "the rational air-conditioned house," but it was one of the clearest expressions of that goal. Air-conditioning experts pushed for a sophisticated balance of architecture and mechanical services. They argued that traditional architectural features that provided ventilation could be eliminated as unnecessary in a mechanized building. At the same time, new features, such as insulation, siting, and landscaping, gained in importance for their role in decreasing heat gain and thus reducing operating costs. Such a balance of features would create a rational design that was affordable from beginning to end.

Industry leaders were clear and outspoken about the kinds of designs that would achieve that goal. For example, generous eaves overhanging the windows, smaller window areas, a reflective color on the roof, reduced wall exposure, and insulation in the ceiling and walls all contributed to a low heat gain. In addition, siting the house so that hot summer sunshine did not hit the windows kept the house cooler. Landscaping to reduce glare and to shade the roof was also helpful. In 1954, when *House Beautiful* hired Henry Wright as "air conditioning and climate control editor," he filled his column with advice on just those features that made air conditioning more efficient and less costly for the consumer to operate.[83]

Unfortunately, the implementation of the rational air-conditioned house depended heavily on the unity of design, construction, and ownership exemplified by the commissioned house. In contrast, tract developments were created by one group and occupied by another. This meant that "consumers" of air conditioning were actually split into two groups, "buyers" and "users." This fragmentation of building practice doomed the holistic vision of air-conditioning experts who advocated equal attention to initial and operating costs. Speculative builders were indeed attracted to novel ideas for reducing construction costs, but they were less concerned with operating efficiency. Air conditioning became affordable when linked to the new tract development, but within this structure the engineering values embodied in the rational air-conditioned house never gained a secure footing.

Speculative builders trying to provide affordable homes adopted few of the innovations calculated to produce the rational air-conditioned home. Tract developers often employed both a design aesthetic and a method of construction unfavorable to economical operation of air conditioning. They engaged in the wholesale clearing of lots to the detriment of landscaping, used identical designs for different exposures, and adopted routine siting of the house upon the lot. The designs sometimes skimped on such relatively costly features as insulation, attic space, and overhanging eaves. Indeed, the new residential architecture often put environmental performance at the bottom of its list of priorities. A noted architectural journal disparaged the new mechanical approach to environmental control: "These new industries with their devices and materials, offer ambrosial hopes for bright, new, indoor tomorrows. But . . . they threaten to turn the building industry inside out simply by proving that any structure can be made livable provided the air conditioner is powerful enough."[84]

The problems of installing air conditioning in such a design did not worry some builders, for the biggest problem associated with such a mismatch was high operating costs, which were shouldered by the new owners. Predictably, the focus upon providing affordable housing through the private sector, drawing upon the tract development strategy, yielded designs that emphasized reduction of the initial cost and neglected the issue of operating costs. The experts' vision of the rational air-conditioned house was shouldered aside by cheaper designs and ignored by the construction industry.

The prevalence of air-conditioned houses with few of the features necessary to reduce heat gain demonstrated that air conditioning was most often installed as a means of achieving affordable home ownership or a new architectural aesthetic freed from the practical considerations of environmental control. The rapid acceptance of central air conditioning among builders was not entirely the result of the air-conditioning industry's efforts to promote the technology as a nearly cost-free luxury. Instead, the peculiarities of tract housing and modern design made air conditioning virtually essential. Tract houses were hot houses. Lightweight construction made them cheap and large picture windows gave them a modern look. An observer noted in 1953 that "central home air conditioning has suddenly begun to boom. Perhaps the most important reason, aside from the recent increase in home ownership, is that today's small house with its sealed picture windows and low roof, is a TV-equipped hotbox that both demands and lends itself economically to a cooling system."[85]

Largely as a consequence of modern design and construction imperatives, then, air conditioning moved quite rapidly from a luxury to a necessity in the building industry. In St. Louis in 1953 the FHA, which underwrote low-cost home loans, still regarded such mechanical equipment as an extra and would not accept it as part of a mortgage.[86] By 1957, however, the FHA had reversed itself and begun to cover the cost of many appliances, including central air conditioning, in "package mortgages."[87] Similarly, private lenders added their voice to the decision-making process. The California Federal Savings and Loan Association of Los Angeles, for example, not only included the cost of air conditioning in its home mortgages, but by 1957 *required* roughing-in for air conditioning in any home costing more than $20,000. Roughing-in included large insulated ducts, adequate space for the equipment, and a return line in homes with slab floors.[88] These additions, which usually ran $250–$300 during construction, cost a fraction of what they would cost if done at a later time.

Such policies were crucially important in promoting the widespread adoption of air conditioning. Institutional approval was an important platform upon which to build public acceptance. By the late 1950s central air conditioning was firmly embedded in new home construction.

THE AIR-CONDITIONED WORKPLACE

The adoption of air conditioning in the office exactly mirrored the pattern of residential adoption. The industry worried about public acceptance of comfort, fought over central versus unit conditioners, and in the end was swept up in the exigencies of commercial design.

At first, the air-conditioning industry analyzed its prospects in terms of individual acceptance of a higher standard of comfort. Promoters of air conditioning for the factory and office fretted over consumers' attitudes about comfort in a business culture that valued hard work and productivity. Carrier

Corporation's postwar attempts to broaden the rationale for factory air conditioning beyond process control to include worker comfort stumbled over the word *comfort*. As late as November 1958 an advertising campaign promoting worker comfort that was slated to run in *Fortune* and *Business Week* drew doubts from company executives. The firm used the term "industrial comfort" for these kinds of systems, but the advertising department was leery of its appeal. One internal critic pointed out that in 1959 American industry was "going to want to get labor to work harder. Management will be 'sweating out' a difficult cost-price squeeze and their thinking will not be to provide 'comfort' for labor. They will expect them to 'sweat' too . . . It seems to me that 'efficiency' is undoubtedly a more appropriate theme for us to use."[89] Carrier Corporation's chief executive officer, Cloud Wampler, was convinced and concluded that "something other than 'comfort' has to be the keystone of this campaign."[90] Eventually, the industrial comfort campaign was abandoned.[91] In the industrial world, at least, *comfort* still carried overtones of ease and relaxation. Such anxieties led some advocates to try to dispel the association of comfort with luxury, by means of careful language: "The word comfort, as here used, does not imply a condition of languorous ease, but rather a composure of body and mind which will increase the alertness and the productivity of the individual by removing distracting and vitiating factors of atmospheric environment."[92]

Office efficiency proved to be easier to sell than industrial efficiency. Ad hoc tests of federal employees in 1946 suggested that typists were more productive in an air-conditioned office. Typists transferred from a regular office to an air-conditioned space increased their output by 24 percent.[93]

In offices, where workers' pace and discipline were not defined by piecework incentives or assembly lines, claims for intangible incentives to greater productivity, such as "heightened morale" and "increased energy," seemed convincing. The reasoning that greater personal comfort would lead to higher worker productivity found more favor among managers of white- collar workers than it had among factory supervisors. Its acceptance as an aid to office efficiency was no doubt helped by the fact that management and office workers often shared the same general work space.

Certainly, the debilitating effects of summer heat were not hard for management to see. Unusually hot summer weather in New York City in 1949 forced many firms to close their offices by early afternoon and send employees home. Even when offices remained open, "efficiency and morale of those employees who stuck to their jobs on hot and muggy days dropped to low levels."[94] Even worse was the heat wave of late June 1952. From New York City westward to Detroit, most offices and factories without cooling released their employees by mid-afternoon. At Coney Island and Rockaway, New York, thousands of people sought relief by sleeping on the beach overnight.[95] The next year a heat wave struck Washington, D.C., in August, sending more than 25,000 federal employees home after lunch one day, and more than 13,000 the next day as well.[96] Indeed, closing the office was a routine response to hot and humid weather, and many managers subscribed to a standard formula for determining when to release employees: When the temperature plus 20 percent of the humidity reached 100 or more, everyone gave up and went home.[97] Such hot spells were only the more dramatic demonstrations of the general malaise that air conditioning could relieve.

As in homes, so in office buildings the high costs of installing central air conditioning in existing structures led to the popularity of window units. In Akron, Ohio, one office building was equipped with fifty-one window units; the air-cooled condensers required no cooling water, the equipment cost one-third less than a large unit, and even installation and maintenance costs were lower. Besides these advantages, each unit was on the power line of the tenant.[98]

If landlords liked the window units, tenants loved them. Tenants in rented office space took their window coolers with them when they moved. Window units became the first choice of the temporary and the mobile. The units provided personal and portable comfort wherever their owners went. Their character as an adjunct to personal comfort was illustrated by the case of federal employees who installed their own window units at work. The federal government argued that since the machines ran on office electricity, all such units became office property, but employees' vigorous protests forced the government to back down.[99] Window air conditioners continued to belong to people rather than buildings.

In Houston, Texas, in 1952, 196 individual offices installed room conditioners, while only 11 office buildings adopted central air-conditioning systems. Despite the overwhelming preponderance of room conditioners in this survey, the total tonnage, or cooling capacity, of the two types of air conditioning was more nearly even: 3,131 tons in room conditioners and 2,327 tons in central-station systems.[100]

Like home construction, new office construction offered a chance to air-condition corporate workspaces at a fraction of the cost of retrofitting an older building. As early as 1950 one air-conditioning journal claimed that "all office buildings that are being erected today, and those still on drafting boards, include air conditioning."[101]

Wampler argued that, more than local weather, a competitive business climate dictated the inclusion of air conditioning in a new office building. He speculated that whenever 20 percent of the office buildings in any one city included air conditioning, the remaining buildings must air-condition to maintain their first-class status. In New York City between 1945 and 1957, seventy-one large buildings, with 23 million square feet of space, were built with air conditioning. The critical 20 percent level was reached in 1953, when renovation was done in twenty-four major buildings.[102] Wampler judged that Philadelphia had also reached the critical level, and by 1955 Philadelphia Electric Company reported that "air conditioning had become an accepted requirement in all new office buildings." In the utility company's opinion, cost had become "secondary to comfort," for one office building that had cost $10 million spent an additional $5 million to add an air-conditioning system.[103] Rising expectations about personal comfort were believed to be a powerful ally for the industry. Wampler told businessmen: "Tomorrow your employees will find non-air conditioned offices unacceptable. The trend is inevitable."[104] In 1957 a survey of 376 companies revealed that 88 percent rated air conditioning the most important item for "office efficiency."[105]

Wampler was wrong in his belief that changing social values were the primary reason for air conditioning's near universal inclusion in new office construction. Even in a new building, air conditioning was an expensive extra, accounting for 16 percent of the total building cost in New York City skyscrapers, the second biggest single item after structural steel.[106] Builders were willing to make that investment because the appearance and popularity of the modern office block demanded more sophisticated mechanical ventilation. Just as postwar home design and construction relied upon air conditioning as an essential element, so did modern office construction.

As urban land became ever more expensive, builders settled on a simple block-shaped building as the most efficient use of land. On the one hand, the sprawling H-, T-, and L-shaped buildings that ensured a high proportion of window exposure to interior space became increasingly expensive as land values rose. On the other hand, however, economical block-shaped buildings left a large number of windowless interior spaces without natural ventilation or light. This interior area, called "deep space," could be used comfortably with the addition of cool fluorescent lighting and air conditioning. Most postwar builders realized that an office building's deep space was now quite profitable.[107] Despite its expense, then, an air-conditioning system facilitated the construction of a less costly building style. The office block was a design dependent upon air conditioning to make it habitable.

The importance of architectural exigencies was revealed in a 1957 survey of fifty power companies, conducted by the Trane Company. Noting that the South exhibited only a slightly higher rate of adoption of air conditioning than other regions, the survey concluded that "outside temperature alone is not responsible for the growth of air conditioning." The report highlighted the developing trend toward air conditioning in areas "not normally regarded as outstanding markets because of their mild climates." The spread of air conditioning in temperate areas was linked to its increasing use in high-rise buildings. Pacific Gas & Electric reported that "three of the newest 8 to 12 story buildings planned for downtown San Francisco will be completely air conditioned, and it appears that there will be a market for commercial air conditioning in our major office buildings from now on." The Seattle Department of Lighting returned a similar account: "Although the Seattle area seldom experiences excessive summer temperatures, there are many cases where the internal heat load of the building makes air conditioning mandatory."[108]

As early as 1929 Willis Carrier had understood the link between new office designs and air conditioning. After contracting for his firm's first multistoried office installation, the Milam Building in San Antonio, Texas, Carrier predicted that mechanical systems would completely supplant window ventilation in the office building. "The office skyscraper of the future may find it preferable to dispense with windows and the street noises and dust which they admit," Carrier argued. "Ventilation in summer and winter can be made much better than outside conditions."[109] Man-made weather would have its own design advantages, such as the elimination of interior courts and areaways that were built to provide window light and ventilation but took up valuable space in congested urban areas. Carrier Corporation's sometime publicist Esten Bolling had elaborated these advantages: "The office building and the hotel of the immediate future must include air conditioning as an essential factor of initial design. The utilization of air conditioning will enable the prescient architect to go about his designing unfettered by the erstwhile necessity of 'ventilating shafts,' 'light wells,' 'outside exposures,' and such considerations."[110]

Architects did appreciate the freedom from the practical considerations of environmental design. Ironically, although air conditioning freed them from the tyranny of window ventilation, it allowed them to indulge in a passion for glass window-walls, provided clients were willing to pay the high cost of cooling. As in the 1950s home, air conditioning underwrote the lavish use of glass in American office blocks. For example, architects Harrison & Albramovitz, Gill & Harrell ran the windows all the way to the ceiling in the United Nations headquarters, and placed a recessed pocket at the top for the venetian blinds to avoid blocking any sunshine. This building was indeed striking, but its profligate use of glass increased its air-conditioning requirements by 50 percent.[111] One air-conditioning expert estimated that for every 100 square feet of unshaded, unfavorably oriented glass in a tall building, an additional ton of air conditioning would be required, at a cost of $300–$800.[112] In 1957 *Architectural Forum* commented that "the architect, with little thought to integration, has led the practitioner of environment control over a dizzying series of jumps— of which the broadest is the ever-growing expanse of heat-entrapping window walls. This has led to the highest costs for comfort in history."[113]

Yet unlike new homes, the modern office building was still produced under the older commercial relationships of custom design, which united technical expertise and ownership in a collaborative venture. The result was that the office building was more likely to embody the ideals of mechanical precision, engineering control, and ideal standards that characterized the earliest air-conditioning installations. More likely to reflect a balancing of long-term and short-term design advantages, the office building combined powerful and sophisticated environmental systems with sealed windows and insulation.

Along with windowless interior spaces and a phenomenally high heat gain, the prevalence of sealed windows was a third major reason why air conditioning rapidly became a necesssary part of the postwar office building. Unopenable windows were a natural component of rational air-conditioning design. They prevented the infiltration of hot air and interference with air-distribution patterns. But like other design elements that complemented air conditioning, the lack of window ventilation made office buildings dependent upon mechanical systems. Air conditioning replaced rather than supplemented older types of ventilation and cooling.

The combination of block design and air conditioning quickly became the norm. In 1955 the federal government gave its imprimatur to the informal standard, when an advisory panel recommended that new government office construction should utilize block-type buildings, "made more feasible than ever by advances in air conditioning and lighting techniques."[114] And in an acknowledgment that the architectural design and the mechanical systems formed an indivisible whole, the panel noted that with block-type buildings and high lighting intensities, air conditioning had in many instances become a "must."[115] Recognizing that notions about the virtue of comfort had little to do with the decision to install air conditioning, the task force reported that "aside from the merits of air conditioning, it is rapidly becoming an accepted necessity."[116]

In light of such recommendations, in 1955 the General Services Administration (GSA), which manages federal office buildings, courthouses, and post offices, adopted a policy that all new buildings be equipped with air conditioning when outside conditions reached sustained periods of 80 degrees effective temperature.[117] This was a decrease from a earlier standard of 84 degrees and effectively moved the boundary from a line just above Washington, D.C., to one that incorporated all but the northern tier of states. Thus, virtually all new government buildings would be air-conditioned. A GSA official acknowledged that "air conditioning has made the 'block' type building possible, just as steel and the elevator made the skyscraper possible. I doubt if we ever build a 'wing' type building again."[118] In 1956 GSA received a $181 million appropriation to air-condition federal buildings, of which $4.5 million was earmarked for Washington, D.C., and $13.6 million for other areas.[119]

AN ACCEPTED NECESSITY

Thus, instituional approval was a driving force in the spread of air conditioning. Central air conditioning was installed in American homes and offices less by the homeowners and corporate staff who used it than by builders and architects. Postwar construction and design used the power of the new technology to create a comfortable environment in an inexpensive but attractive building. Both the small picture-window house and the glass- fronted office block were heavily dependent upon air conditioning to make their economical designs habitable. The decision by the FHA to include air conditioning as part of government-insured home mortgages and the GSA policy of incorporating air conditioning in new federal office buildings were acknowledgments that the technology was an integral part of postwar building design. The experiments in air-conditioned homes and offices, begun by 1952, were thoroughly established by 1957. Those policies marked air conditioning's institutional acceptance and provided a framework within which individual decisions about personal comfort were often made.

While the makers of window air-conditioning units still worked within a system that bundled together mass production, lower prices, and consumer acceptance as a comprehensive business strategy, manufacturers of central air conditioning also made their product competitive by following that prescription and more. Indeed, central air conditioning was increasingly produced as a packaged unit in standard sizes at lower cost, and industry leaders were vocal about the benefits of air-conditioned homes and offices. They believed it would be impossible to gain consumer acceptance without reducing both the price of their equipment and the cost of operating it. But air conditioning first gained its new status as a necessity rather than a luxury among builders and architects who used it to underwrite modern design and construction. Air conditioning was essential because the modern house and office were redesigned around its mechanical systems and often dependent upon them. This disjunction between the buyers of air-conditioning systems and their ultimate users encouraged the separate consideration of initial costs and operating expenses. Thus, although the manufacturers of central air conditioning triumphed by once again integrating the technology and the building, their vision of a rational air-conditioned house was pushed aside by the forces of the marketplace.

CHAPTER SEVEN

Consumers

and Air Conditioning

*T*he advocates of central air-conditioning systems hoped to make the technology competitive by linking it to the revolution in mass-produced tract houses that characterized the government's push for affordable home ownership. Yet the approval of architects, builders, and bankers between 1946 and 1960 was only one element in its eventual widespread adoption. Institutional acceptance of residential air conditioning was necessary but not sufficient to convince homeowners of its advantages. In addition, the success of residential cooling was assured by a cascade of individual decisions to buy air-conditioned comfort. That consumer approval was hard won, for homeowners were slow to put aside hot-weather traditions in favor of mechanization, and owners of air conditioning were reluctant to accept authoritative prescriptions about its use. Once the decision was made, however, its impact upon social patterns and public services was enormous. By the mid-1960s, the demand for vast amounts of cheap energy to run both air conditioners and a panoply of other consumer gadgets strained the capacity of the nation's electrical network. The sum total of consumer decisions thus created a huge public policy problem.

THE SUCCESS OF CENTRAL SYSTEMS

Central air conditioning became an integral and necessary part of the construction industry because of lightweight construction, hot exposures, large windows, sealed windows, and no windows. The adoption of air conditioning as an adjunct to architectural design was most common in the modern office block, yet the suburban tract home had a similar built-in need for mechanical environmental control. However, the effectiveness of this top-down process of selling—to construction professionals and only indirectly to the ultimate users—is harder to evaluate in the case of residential building.

The Bureau of the Census first surveyed residential air conditioning in 1960 and was surprised at its own findings; the bureau reported that "although regarded as a luxury item not many years ago, air conditioning was reported in 1960 for approximately 6.5 million occupied housing units (1 in 8)." Despite the evident popularity of air conditioning, the total number of central systems was still small—996,000 units or approximately 2 percent of the total housing stock. Window units outnumbered central systems five to one.[1]

However, national statistics obscure some interesting regional patterns. In some localities central systems rather than window units had become the norm. In metropolitan Phoenix and Tucson, Arizona; Las Vegas, Nevada; and Bakersfield, California, more households relied upon central air-conditioning systems than on window air conditioners.[2] Residents of those areas took to heart the idea of building in air conditioning from the start. Not only was central air conditioning competitive with its rival, it was becoming common among homeowners. In Phoenix and Las Vegas nearly a quarter of all households had such a system.[3] The next census in 1970 revealed that Dallas and Austin, Texas; Washington, D.C.; and Fresno, California, had joined those cities that favored central systems. By then, the percentage of centrally air-conditioned houses ran even higher: 42 percent in Dallas and Austin, 40 percent in Washington, and 29 percent in Fresno. To achieve such high levels when measured against the total housing stock, a very high percentage of new homes must have incorporated air conditioning. The pervasiveness of central systems suggests that there was nothing elitist about their ownership in these communities.

Many central systems were a product of the political economy of speculative building. A survey conducted for Carrier Corporation in 1959 by Gallup and Robinson suggested that 46 percent of centrally air-conditioned houses were constructed by speculative builders. Many builders believed that air conditioning helped them sell houses; from ten promotional features designed to appeal to homebuyers, builders ranked air conditioning second, after built-in kitchens. And while a majority of builders denied that construction costs could be reduced by designing a house around air conditioning or reported that they did not know how to do that, the remainder stated their belief that money could be saved by using fewer windows or sealed windows, using proper ducts, and lowering the height of ceilings. Indeed, 7 percent of homeowners acknowledged that the primary reason they had air conditioning was because the "house was built for it."[4]

The efforts to make central residential air conditioning a mass market included rhetoric from manufacturers about making the whole house absorb the cost, the implementation of that strategy by tract developers, and supportive policies among financial institutions. But in spite of these efforts, in 1960 central residential air conditioning remained uncommon on the national scene.

The cost of air conditioning remained an important factor in consumer decision making, and income was an important predictor of ownership even at the end of the decade. National surveys indicated that the majority of owners were college-educated professionals, managers, and business owners, with an average income of $11,508, in contrast to a national average of $5,083.[5] Air-conditioned houses reflected that prosperity; Gallup and Robinson suggested that a third of the centrally air-conditioned homes were worth $30,000 or more. Consumers, builders, architects, real estate agents, and bankers all concurred that the expense of an air-conditioning system remained the biggest obstacle to its widespread adoption.[6]

The wisdom of industry efforts to make the whole house absorb the cost of the air conditioning is reflected in Gallup and Robinson's findings that if an air-conditioned house cost no more than an ordinary house, 86 percent of consumers would buy one. Indeed, only a small number of people seemed to actively dislike air conditioning; a recalcitrant 11 percent of homeowners reported that they deliberately avoided public air-conditioned places like theaters and restaurants. The most frequent complaints against such places were that the contrast in temperature between the inside and outside was too severe, that the inside was too cool, and that one or the other would promote more frequent colds. Although air conditioning won frequent approval from the general public, as the price of air conditioning rose, consumers rapidly lost interest in buying a system: 63 percent claimed they would pay $200, 39 percent would spend $400, 15 percent would part with $700, and only 7 percent were willing to hand over $1,000.[7]

PUBLIC ACCEPTANCE

However, the price of an air-conditioning system was not the only obstacle to ownership. Public acceptance remained an important factor in air-conditioning sales. The postwar planners of the air-conditioning industry wanted to convince the public that air conditioning was the answer to hot weather, but by 1960 they were still not really sure they had succeeded. After surveying the general public, Gallup and Robinson told Carrier Corporation, "Heat, humidity, dirty air, appear accepted as unpleasant facts of life. Many people appear to accept these problems as inevitable, endurance becoming habitual." Although 66 percent of those polled said that they believed the advantages of air conditioning were greater than the disadvantages, only 1 percent of families reported that they were saving for either central air conditioning or a room conditioner as their next major purchase.[8]

Although consumers disliked heat, humidity, dirt, and pollen, few sought air conditioning as the solution to these problems. Instead, generations of Americans continued to use traditional ways to beat the heat, short of mechanical air-conditioning systems. From air domes to swimming holes, outdoor amusements ranked high in hot weather. For the more sedentary, simply sitting on the porch or stoop was a favorite pastime. When it was time to go to bed, some favored the sleeping porch. Any method for catching a little breeze made high temperatures more tolerable.

Folkways of dealing with the heat often involved getting out of the house. If windows are kept closed after the cool morning hours, the indoor temperature of most houses will lag the outdoor temperature by about three hours. So while the house may provide a shady retreat for part of the day, by early evening the outdoors has begun to cool down while the house reflects the temperature of the hottest part of the day. Abandoning the house for a splash at the beach during the afternoon or for the breezy porch, patio, or stoop in the early evening was a time-honored strategy for avoiding the heat.

Ironically, the natural climate provided the antidote to its own excesses. So it is not surprising that the idle and prosperous sought a more ideal natural climate for summer relief. Often wives and children left for a temperate resort, while the family breadwinner remained in the city and became a "summer bachelor." Hot-weather cities like Phoenix were not exempt from this pattern. Longtime Phoenix resident George H. N. Luhrs Jr. savored boyhood memories of summertime dips in the city's canals and sales of lemonade at a stand in front of the saloon. When these amusements were not enough, the Luhrs family vacationed at various California beach resorts that regularly advertised in the Phoenix newspaper.[9] McKean's Model Laundry and Dry Cleaners counseled readers of the *Phoenix Gazette:* "When wife's away all I say is Bachelor Service . . . If you're a summer widower banish the worry over clean linens and neat appearance!"[10]

The persistence of such patterns of outdoor living was obvious to the air-conditioning industry; indeed, the tradition appeared to grow stronger in the postwar era as more and more families moved to the suburbs. A study conducted for Carrier Corporation in 1948 noted glumly that suburban homeowners considered their location " 'at least 10 degrees cooler' than the city" and that "they tend to live more and more in the outdoors in the summertime, eating, playing and sleeping outdoors and on porches." Particularly for the "bread-winner," home only on the weekends and the cool of the evening, the company feared that an attic fan would be deemed sufficient for comfort.[11]

Clearly, the acceptance of air conditioning turned on a transformation of values. Consumers had to accept air conditioning as a necessity, just as building professionals did. The 1950s were a time of transition in attitudes. Air-conditioning pioneer Walter Fleisher pointed out that

> not many years ago, we would not have considered spending large sums of money to provide cooling in ordinary places of business or the home. Of course, we did, if we were able, move to the seashore, the mountains, or the country for a short time with the idea of enjoying relief from summer heat. But today we have become so accustomed to luxury that we consider it our inalienable right to have every type of comfort that our creative technology has developed and among these is artificial cooling.[12]

When the Central Arizona Light and Power Company promoted air conditioning, it targeted the Phoenix tradition of a summer getaway: "Take a vacation from the heat in your own home with modern air conditioning and room cooling. You can manufacture your own weather. Even the respite of a brief vacation seems only to make the rest of the season more a period of wretched endurance."[13] For every industry analyst agonizing over the persistence of old values, there was a social critic decrying the power of the new. The *New Yorker* insisted that "we have long argued that in our youth, summers in New York were supportable without air from air-conditioning." However, the editors recognized that times were changing: "The dodges for coping with the heat that New Yorkers learned in three centuries of summers have become superfluous . . . The long drink is an irrelevancy; if you arrive in a bar, after a few steps in the street, longing for a Tom Collins, half a minute of the temperature inside influences you to change to a hot toddy. Cold foods lose their charm as quickly; at the first blast of frozen air, the customer decides to stick to steak."[14]

The air-conditioning industry directly addressed the social rituals of keeping cool. Industry leaders hoped to convince consumers that new family patterns were one of the advantages of the new technology. Thus they closely watched the Austin Air Conditioned Village for an abandonment of traditional hot-weather strategies. And, indeed, researchers reported that families in these air-conditioned houses cooked more, baked more, ate heavier foods, drank more warm drinks, attended fewer movies, organized fewer picnics, stayed home more, entertained at home more, and kept the children inside more.[15] According to the director of the Good Housekeeping Institute, keeping the children at home was counted by consumers as one of the most positive aspects of air conditioning.[16]

MAN-MADE WEATHER

To convince the public, industry advocates turned again to an established theme, the superiority of man-made weather over natural climate. The ability to create an artificial environment indoors raised the standards by which to judge natural climate and comfort. Using U.S. Weather Bureau data, anyone could count the precise number of days in which the outdoor conditions were within the experimentally defined comfort zone; for example, *Life* pointed out that St. Louis enjoyed only forty-five perfect days a year.[17] Few climates could stand the comparison to the ideal that emerged from laboratory definitions of comfort.

The comparison of the natural climate to a quantitative ideal led to higher standards. Such rising expectations can be clearly seen in the changing attitudes toward California's climate. One early resident of Los Angeles was ecstatic in his praise of the state's weather, asking, "In this land of perpetual summer and 'sweet do nothing,' why should anybody go into the house except to eat and sleep?"[18] But later advocates of air conditioning argued that "even California's salubrious climate must be capable of improvement and correction."[19] Such a change in perceptions about the relative merits of natural climate and man-made weather did not come inevitably or unassisted.

The deficiencies of specific cities became apparent to all in June 1959 when the U.S. Weather Bureau began to release an index of human discomfort experienced as a result of the combined effects of temperature and humidity. The bureau named this the Discomfort Index.[20] It was the outgrowth of decades of research into the relationship of temperature and humidity, beginning with the comfort zone research of the 1920s. Such popularization of these concepts represented the spreading influence of quantifying comfort. However, the Discomfort Index met with strong opposition, as cities such as New York City and Miami strove to protect their tourist industries from the implied criticism.

Within two weeks of its introduction, it was given the much more neutral name of Temperature-Humidity Index.[21]

How readily did consumers embrace the ideal of man-made weather? Man-made weather was a comprehensive, rational vision based on the assumptions that comfort should be quantitatively defined, that mechanical systems should replace natural ventilation, that these mechanical systems should create an indoor environment as complete in its essentials as outdoor weather, and that such systems should be designed and run by technical authority. But individual consumers instead were erratic and pragmatic. The ways in which they used their air-conditioning systems suggest that they incorporated air conditioning into their lives on their own terms, independent of ideas from the industry.

The industry never altogether succeeded in convincing the public that air conditioning was indoor climate with a range of functions similar to natural climate. From the time when movie theaters first advertised by hanging icicles from the marquee to the end of the 1950s, the general public always equated air conditioning with cooling. Indeed, as the maturing technology abandoned the air washer and its plumbing connections, humidity control often became less precise; summer air conditioning was more likely to settle for whatever dehumidification was a consequence of a decrease in the temperature. Similarly, the smaller equipment was less likely to "wash" the air and more likely to rely upon filters to clean the air. But central residential air conditioning continued to supply some variation of its original four functions—regulating temperature and humidity, cleaning and circulating the air—as essential elements of climate. Nevertheless, the public image of air conditioning solidified independent of this shifting technological reality. Eighty-six percent of the respondents to the 1959 Gallup and Robinson survey identified air conditioning with cooling and comfort, while only 10 percent identified control of hay fever, asthma, pollen, or humidity as a reason to air-condition one's home.[22]

Other elements of the man-made weather ideal fared only slightly better than the notion that air conditioning was analogous to natural climate. One of the hallmarks of man-made weather was maintaining comfort levels within experimentally defined parameters, yet engineer Charles Leopold noted that American homeowners in general maintained conditions at their own preferred temperature "rather than the statistically found optimum."[23] His observations were borne out by studies of the residents of the Austin Air Conditioned Village; residents paid little attention to laboratory-defined standards for perfect indoor comfort. All villagers set their thermostats below 80 degrees (one at 70 degrees, four at 72, two at 75, three at 76, two at 77, and six at 78).[24]

In addition, many Americans did not welcome the substitution of mechanical systems for natural ventilation that was another part of the man-made weather ideal. The sealed window became one of the least favorite aspects of the new home and office designs. Indeed, at the insistence of the client, CEC had designed its first multistory office installation, the Milam Building in San Antonio, Texas, so that any of the forty-six tenants could open their windows without affecting distribution.[25] Tenants were especially critical of the sealed windows of later office designs whenever air-conditioning systems failed. One maintenance engineer related his experience when the air-conditioning system faltered on the hottest day of summer. Complaints about conditions started by 11 A.M., but the hapless engineer could not pinpoint the source of the trouble. He remembered that "as people got back from lunch, they were boiling in more ways than one. Coming from the hot street, instead of comfort, they found it was hotter than ever. They got irritable and mad at everybody, and the phone calls got more belligerent. Besides me, the chief target was the guy who designed the building with sealed windows."[26]

Similarly, few consumers seemed to favor the sealed windows of the rational air-conditioned house. *Fortune* reported that "some persistent element in people makes them want to open windows on a beautiful morning. Women, it seems, are especially stubborn about this."[27] Although consumers continued to insist on windows that could be opened, they showed some willingness to follow the industry's advice to keep those windows shut. In the Austin Air Conditioned Village, eighteen of the twenty-one families kept their windows closed 24 hours a day.[28] But even closed windows met with some resistance; 6 percent of consumers surveyed by Gallup and Robinson listed a closed house as a reason not to air-condition.[29] In the Austin Air Conditioned Village, three families complained of a lack of fresh air, and members of one family preferred to sit on the patio in the evenings because the air was "stale" indoors.[30]

If exacting technical experts required air-conditioning systems to produce conditions within the comfort zone, they presumed that consumers would operate the installation accordingly. In the case of systems sized by the 24-hour method, the link between design and operation was crucial; at extreme temperatures those systems would maintain comfortable conditions only if run continuously. Accordingly, many consumers relied upon the sophistication of the design and controls to establish and maintain the proper environmental conditions. According to Gallup and Robinson, roughly half the homeowners surveyed were passive operators who let their systems run continuously; they used their air-conditioning system for four or five months of the year (49%), all summer long (51%), or 24 hours a day (41%), and seldom adjusted the settings. Indeed, when ten housing experts from the Soviet Union toured the Austin Air Conditioned Village at the invitation of the NAHB, they agreed that the " 'greatest merit' of the village's air conditioning is the fact that it requires no technical knowledge on the part of home owners."[31] Nonetheless, homeowners, like factory workers and schoolteachers before them, could be problematic operators of air-conditioning systems. Although many users let thermostats regulate their systems, a sizable minority were fiddlers. A third of owners turned the air conditioning on only during the hottest days (36%), changed the setting at night (36%), or changed it according to changes in the weather (18%).[32]

AIR-CONDITIONED LIVING

Many consumers resisted the technical authority that was an integral part of scientific definitions of comfort, mechanically dependent buildings, rationally designed and operated systems, and comprehensive environmental models. Many did so simply because no single set of design assumptions could accommodate the diversity of family patterns and individual needs. One husband and wife in the Austin Air Conditioned Village both worked outside the home. They reported that they turned the air conditioning off in the morning and turned it on again upon coming home. Researchers pointed out that this pattern of operation did not really establish "comfort conditions" until after midnight.[33] Yet the couple were satisfied. They were among the owners who operated their systems to fit individual needs not envisioned by designers and manufacturers.

Indeed, the industry sometimes designed residential systems according to an idealized notion of family life. The home was envisioned as a place for relaxation, entertainment, and socializing. It was a place where the husband habitually put up his feet and read the newspaper, the family periodically gathered around the television set, and the neighbors occasionally came over for drinks. Defined in opposition to the workplace, home was expected to be a comfortable retreat and a place of leisure.

But if the home was imagined to be a man's castle, it was in reality a woman's workplace. Accordingly, the industry promoted the advantages of the air-conditioned house to housewives. So proponents argued that the closed windows of the air-conditioned house decreased the amount of dirt that blew in from outside. Indeed, women in the Austin Air Conditioned Village reported less dirt and dust in the house.[34] Willie Mae Rogers, director of the Good

Housekeeping Institute, reported that, on the whole, women liked the comfort of air conditioning and "the greater cleanliness of the home."[35]

House Beautiful touted that advantage, reminding readers that because "windows aren't opened, city dirt and noises are kept out." Like many domestic technologies, air conditioning was viewed by the magazine as a means of elevating the standards of style and cleanliness in the home rather than simply a way of saving time on housecleaning chores. The "perfect" bedroom in its air-conditioned house included such luxuries as a white cotton rug and sheet-white draperies, which the editor conceded would have been "impossible" without air conditioning.[36] The addition of such high-maintenance items as white rugs meant that the housewife who adopted this approach to air-conditioned living would not spend less time doing housework despite the decrease in dust and dirt. Women in the Austin Air Conditioned Village reported spending more hours on housework rather than less. They baked and cooked more than in previous summers, serving their families more hot food.[37] Clearly, the village housewives pursued a course that placed family welfare ahead of personal leisure.[38]

Although promoting the advantages of air-conditioned homes to housewives, the industry was slower to come to terms with more problematic aspects of women's housework. Especially troublesome for the air-conditioning system was the heat load of domestic appliances. Cooking, baking, ironing, dishwashing, and clothes drying were the greatest sources of internal heat, but even electric appliances such as vacuum cleaners added to the total heat that resulted from housework. Some installations simply ignored the whole problem. In one of the homes in the Austin Air Conditioned Village, for example, the occupants were troubled by excessive humidity from a clothes dryer that was unvented. Most, but not all, of the homes in the village had kitchen ventilating fans to exhaust the heat of cooking. In only a few instances, however, was any provision made for the introduction of outdoor air equivalent to the amount exhausted.[39] Both failures seem to reflect a certain disregard for the exact nature of women's activities in the home.

This is not surprising, for in American society in general, women's unpaid labor in the home was often invisible. In the 1950s, in particular, the social conservatism that pushed women out of wartime jobs and back into the home promoted a view of women's domestic work as affective rather than physical. Housewives were acknowledged to sweep and mop the floors, but their important work was nurturing children, soothing familial tensions, raising cultural values, and maintaining aesthetic standards. As long as air conditioning's special appeal to women was seen as a promise to stretch the interval between episodes of redecorating, it was unlikely that anyone would engage the conflict between summer cooling and the hot kitchen stove.

The physicality of housework and the heat load of domestic appliances intensified rather than declined in the postwar period. Some tasks, such as laundry, which had formerly been taken out of the home and commercialized, now returned to the housewife; commercial laundries peaked in the 1920s and then declined as domestic washing machines undercut their market.[40] In addition, the consumption of electric energy in the home traced a steep upward curve. Just as electrification and increased machinery speeds intensified the heat load of the factory, the growing electrification of the home created a hotter house. Air conditioning needed to address not only the traditional sources of heat in housework but an ever increasing energy consumption. *Fortune* magazine observed that "the American likes his home brilliantly lit, of course, and he has a passion for gadgets that freeze, defrost, mix, blend, toast, roast, iron, sew, wash, dry, open his garage door, trim his hedge, entertain him with sounds and pictures, heat his house in winter, and—above all—cool it in summer."[41] *Fortune* was right that air conditioning itself had become part and parcel of America's passion for electrical household gadgets.

Air conditioning in the 1950s and 1960s traveled a bumpy road. Although makers of central air conditioning proposed a vision of perfect indoor climate, high standards of performance, ideal operating procedures, and a new way of living, consumers seemed to embrace the technology but not necessarily the industry's prescriptions. Some consumers found that package of equipment and ideas immensely appealing. They relished manufacturing ideal weather in their homes everyday and reveled in the escape from hot-weather restrictions on their lifestyle. Others, however, were less exacting in their expectations. They failed to find their own lifestyles reflected in idealized notions about family life, clung to the social rituals of hot weather, and stoutly maintained their rights of control. Yet even this second group increasingly installed and used the new systems. They were not so much holdouts against technology as "irrational" users. They were highly logical in their independent determination of goals and standards of performance, and in the ways they bent the technology if not the marketplace to their individual needs. Theirs was often an ethic of consumption. However, to define and configure a technology solely in terms of consumer desires and the marketplace, however, was as problematic as to bend human possibilities around the requirements of the machine. For there continued to be an engineering rationality that was costly to overlook—a fact that the energy crisis of the 1970s starkly revealed.

EVAPORATIVE COOLERS

The acceptance of air-conditioned living came more easily to some communities than to others. Washington, D.C., embraced the new technology of comfort air conditioning in its early years, but so many of the city's initial installations were publicly financed that its experience was atypical. A scan of the census data reveals that the hot, dry western cities led the postwar boom: first, Phoenix, Tucson, Las Vegas, and Bakersfield, followed by Dallas, Austin, and Fresno. The Southwest and California's Central Valley were exceptional in their acceptance of central air conditioning. This is one of the first geographic, rather than economic, patterns to emerge, for previously income had been the dominant predictor of ownership of air-conditioning systems.

At first glance, this pattern is surprising. Air conditioning began its commercial development primarily as humidity control, and primarily for industrial applications. With the development of comfort air conditioning, experimental science and bypass distribution made humidity control a necessary and economical aspect of comfort. Again, when the industry moved into the residential mass market, its insistence on a full range of functions, including humidity control, was one mechanism for discriminating between competitors. Finally, humidity control remained central to the vision of air conditioning as artificial climate. With all the importance that the industry continued to attach to this aspect of the technology, why did consumers in the arid West, who needed humidity control the least, respond first to the promise of air-conditioned living?

Any answer must surely incorporate the truism that consumers had always thought of air conditioning as cooling, despite continued attempts at public education to promote the idea of air conditioning as artificial climate. Whether theater managers led or followed the public is this respect is unclear, yet from its first public appearance, air conditioning was prized for its cool temperatures. A second factor is the long tradition of mechanical cooling that preceded the postwar air-conditioning boom. The hot, dry climates of the West were perfect for evaporative coolers, and these cheap and effective machines let westerners become accustomed to mechanical cooling at home. The early widespread use of evaporative coolers appears to have paved the way for the adoption of refrigerated air-conditioning systems of all kinds.

Direct evaporative cooling was common in the dry western states. Often called "swamp coolers," these devices were generally quite simple: An electric fan blew air through dampened pads or screens, and the subsequent evaporation cooled the air. In areas with low humidity, the evaporative cooler could reduce the temperature 20–40 degrees or more.[42] Phoenix was an early center for their production and use. By 1939 an estimated 10,000 swamp coolers were in use in Arizona, outnumbering all other types ten to one.[43] As simple in construction as in principle, evaporative coolers were long the object of do-it-yourselfers. In the 1930s two University of Arizona professors, Martin Thornburg and Paul Thornburg, conducted experiments to find optimum designs for the humble swamp cooler, and their mimeographed construction plans circulated throughout the state, probably through the agricultural extension service. Their work was later published as *Cooling for the Arizona Home* (1939). In that year the Thornburgs estimated that a direct evaporative cooler cost $10 to make at home and 1 cent per hour to operate.[44]

Commercial production of evaporative coolers began about 1936 and was centered in Phoenix. The five largest Phoenix manufacturers—Palmer Manufacturing, International Metal Products Company, Polaraire Manufacturing Company, and Mountainaire Manufacturing Company—together grossed $15 million annually by 1951.[45] Other companies were established in Los Angeles and Dallas.[46] By the early 1950s more than 90 percent of Arizona homes had evaporative coolers of some type.[47]

Fresno was another community that adopted the evaporative cooler to take the edge off its hot, dry climate. Located in California's San Joaquin Valley, Fresno became a national center for table grapes and raisins, thanks to its alluvial soil irrigated by water from the nearby Sierra Nevada mountains. Its hottest recorded summer was in 1889, when temperatures surpassed 100 degrees for eighty-two days, and its coolest came in 1975, when the city had only seventeen days above the 100-degree mark.[48]

Such weather contributed to the appearance of the social rituals of hot weather. One long-time resident recalled the white flannels, straw hats, and bonnets of his uncle's generation in 1870, and the hand fans that merchants gave away to customers.[49] Those who could left for cooler climates during the heat. The local paper recorded that "while Dad might be in town tending to business, Mom and the kids were in the mountains or headed for the coast."[50] Wealthy valley families either vacationed at Huntington and Shaver Lakes or left for Monterrey, Carmel, or Santa Cruz.[51] Many of these hot-weather strategies persisted through the postwar era in populist forms, as adolescents continued to cool off by swimming in the irrigation ditches and families escaped in campers to Pismo Beach and Morro Bay.

Those social patterns were soon augmented by mechanical cooling. Affluent Fresno families, like those in other parts of the country, installed their first refrigerated air-conditioning systems in the 1930s. Ted Goth installed what he thought was the first such system in the Samuel Holland residence on Huntington Boulevard in 1932.[52] But even the more modest homes in Fresno could afford evaporative cooling. In those same years, homemade swamp coolers began to appear in Fresno houses, where they were often mounted in the window with a common green lawn hose providing a trickle of water through its excelsior pads and leaving a muddy puddle under the housing.[53] Commercial units were installed in the Fresno County Courthouse at about the same time.[54] By 1953 an estimated 10,000 evaporative coolers were in use in Fresno.[55] Simple and cheap to operate, evaporative coolers nearly sold themselves.[56]

ENERGY CONSUMPTION

While central air conditioning needed to be built into a new home to be affordable, window air conditioners could be added to nearly every building. Adaptable, affordable, and cool, they formed an advance guard for the increasing popularity of air conditioning. By 1960 there were about 6.5 million air-conditioning units of all kinds in use, and by 1970 there were more than 24 million, nearly 17 million of which were room units.[57]

Although air conditioning struggled with the rising heat load of the modern home created by Americans' passion for electrical gadgets, air conditioning was itself part of that obsession. Perhaps more than any other appliance, air-conditioning systems propelled energy consumption to record highs. As early as the 1930s utility companies had recognized air conditioning's potential to provide a summer load to offset winter heating peaks. They had promoted it aggressively with merchandising campaigns. But the reality of widespread adoption far surpassed the utilities' expectations. As early as 1957 Commonwealth Edison Company of Chicago reported that peak energy consumption in August exceeded the winter high in December.[58] *Business Week* reported that "the tremendous growth of air conditioning, largely the result of utilities' advertising, came as a surprise to some companies." The Federal Power Commission also was caught off guard by air conditioning's popularity; its 1964 estimate for a peak load of 494 million kilowatts in 1980 was revised upward by 60 million kilowatts only five years later, due to the growth of air conditioning and space heating. By 1969 some power companies had discovered that 35 percent of their peak demand was due to air conditioning; by then, utilities in the Northeast no longer advertised air conditioning to build loads. Besides, hot weather alone was enough to spur sales of window units. During one long heat wave in the Midwest in the summer of 1969, the Sears Company in Milwaukee sold 7,000 air conditioners in one day.[59]

The first indication that the explosive growth in residential power consumption might present problems for utility systems came in July 1966. A prolonged heat wave that hit the eastern two-thirds of the nation in mid-July 1966 created trouble for several major cities. In New York City it became so hot that the 207th Street Bridge over the Harlem River between the Bronx and Manhattan jammed open when the metal expanded and had to be hosed down to reduce the swelling.[60] Automobiles overheated and stalled in the city's tunnels and overheated drivers responded with a furious chorus of horns.[61] Consolidated Edison assured customers that there was no danger of a widespread blackout, but intensive use of air conditioners caused a series of more limited neighborhood blackouts all during the heat wave.[62] In St. Louis, where the All-Stars Game was being held, Stan Musial looked at the St. Louis Arch and quipped that "all the cool air we'll have this summer will come through the arch as soon as we turn the switch." Temperatures in St. Louis rose to 104 degrees that day, while inside the stadium someone recorded 116 degrees.[63] Unbelievably, the next day brought a high of 106 degrees, and a series of six days over 100 degrees drove the local utility, Union Electric, to desperation. "Intensive use of air-conditioning and other cooling equipment has pushed us to the limit of our capacity," a Union Electric official conceded.[64] Once so committed to the promotion of air conditioning as a summertime load that it had planned on producing its own residential units, Union Electric was now overwhelmed. The company began to cut off power to neighborhoods on a rotating basis in an attempt to avert a major power failure.[65] The city's early adoption of air conditioning now seemed a mixed blessing.[66]

If air conditioning was a bane to the utilities in this crisis, it was a boon to the consumers. Those who lacked it suffered from the heat wave most severely. The death toll in St. Louis rose to seventy before it was over, eight times the normal level of mortality. St. Louis coroner Helen L. Taylor reported that "many of them are from poor areas and you can imagine their living conditions—no fans or air conditioning."[67]

Indeed, low-income groups in all the major cities suffered. In New York City residents of low-cost housing were not allowed to install their own air-conditioning units because the wiring was inadequate.[68] To the extent that African Americans were part of the urban poor, they took the brunt of the heat wave. Census figures show that when the national average was around 12 percent, fewer than 4 percent of households headed by blacks had any kind of air conditioning.[69] In Chicago rioting broke out when police officers turned off fire hydrants that African American children were using to relieve the 98-degree heat. Martin Luther King Jr. intervened to secure a pledge from Mayor Richard Daley that the city would put sprinklers on fire hydrants so that ghetto children could cool off on hot summer days.[70]

ENERGY CRISIS

By 1966 utilities were overwhelmed by hot-weather demand, and the oil crises of the 1970s did little to relieve the situation. By 1979 concern about energy consumption had escalated beyond company strategy or consumer affordability and had become an issue of national public policy. In July of that year President Jimmy Carter declared an energy emergency and ordered all air conditioners in public and commercial buildings set no cooler than 78 degrees.

But Carter's presidential order did not apply to private homes. To some experts, residential air conditioning was equally important in energy conservation, even though the sanctity of the private home resisted public regulation. Many felt that energy conservation efforts were hampered by the legacy of three decades of building practice that had placed affordable home ownership above environmentally conscious design. Customers installed window air conditioners in older homes that were never intended to accommodate them, and builders placed central air-conditioning systems in those "TV-equipped hot boxes." The consequences of that disjuncture between environmentally conscious design and mechanical systems were higher consumer costs, overtaxed electrical networks, and a national dependency on imported oil.

FRESNO'S RETROFIT CAMPAIGN

One of the communities that attempted to address energy usage on the local level was Fresno. Certainly the high percentage of Fresno homes with air conditioning accounted for a large part of summertime peak energy usage. Although 1975 was the coolest summer recorded up to that time, Pacific Gas and Electric Company (PG&E) reported a new high in energy consumption, and a company official admitted that "the surge in air conditioner use was one of the main reasons for the record."[71] By 1979 the utility reported that isolated cases of "brown-outs" or "dim-outs"—losses of power caused by overloaded transformers—occurred every summer, and once again pointed to air conditioning as one of the causes.[72]

PG&E clearly considered postwar construction and the lack of attention to environmental performance as a contributing factor to the amount of energy required to cool area homes. It estimated that fewer than 50 percent of homes had sufficient insulation. A Fresno insulation company reported that the average Fresno home had light insulation in the ceiling and that many had no insulation in the walls.[73] When energy had been cheap and plentiful, that had made little difference; residents simply relied upon their air conditioning to make houses comfortable during the summer. By the mid-1970s, however, that solution was increasingly costly to both the consumer and the utility companies.[74]

In an attempt to rectify the situation, municipal authorities in 1980 proposed bringing the insulation in all Fresno homes up to a higher standard for greater energy conservation. Under this plan, owners would be required to retrofit their homes with adequate insulation before they could sell them. Traditionally, municipal officials had consulted informally with representatives of the construction industry about proposed changes.[75] In this case, they seem to have failed in an effort to build consensus. A storm of protest greeted the retrofit proposal. Homeowners worried about the burden of the expense, especially for older citizens, and protested the intrusion of government on private matters.[76] Builders joined in the criticism. The retrofit effort was dropped.

Fresno was a clear example of a national problem. The ever increasing energy consumption in American homes strained the utilities' capacity to supply enough energy for the needs of the public and regularly pushed the power grid beyond its limits during peak demand. After the mid-1950s peak demand occurred during the summer when consumers turned on their air conditioning. Although some American homeowners criticized, modified, or subverted the industry's vision of air conditioning, they nevertheless adopted the technology with a speed that surprised even its advocates. The energy crisis was more visible in Fresno than in some other communities because of its long tradition of mechanical cooling and its high percentage of air-conditioned homes. But the problem was a national one. In the drive for energy conservation, public officials could not extend their authority into the private home to regulate how consumers set their thermostats or operated their equipment. Nor did rhetoric seem to make a difference. Engineers had been trying to tell consumers about the optimum setting of their thermostats for decades with little apparent success.

In many cases it would not have made a difference. Part of the energy problem was structural—literally built into postwar homes that put modern design and affordable homeownership at the top of the list of priorities. The industry's vision of a rational air-conditioned house had never made much headway in the political economy of speculative tract development. Thus, Fresno's retrofit proposal seemed to be the most effective measure of those that were politically feasible. By integrating the house and its mechanical services, city authorities attempted to reconcile personal freedom with public interests. Unfortunately, they were unable to extract enough concessions from private interests to build a consensus behind their proposal.

CONSUMER ACCEPTANCE

Consumer acceptance of air conditioning was a slower process than its prevalence today might lead us to expect. Consumers were not necessarily anxious to give up the distinct pattern of hot-weather culture—the summer clothes, food, drinks, relaxation, entertainment, and travel—and some interleaved the new technology with the older traditions rather than substituting an entirely new way of life. With a public not entirely convinced of its necessity, income rather than climate long remained the best predictor of ownership. Only in a few cases did regions embrace air-conditioned living, and at a first glance, those were not the most likely. The distinct pattern of consumer acceptance is one more indication that the vision and understanding of the technology that animated air-conditioning engineers—air conditioning as man-made weather—failed to reach or convince a vast number of consumers. These users bought and operated their systems on their own terms.

Conclusion

Air conditioning provides a portrait of an industry that successfully made the transition from custom design to mass production. Although mass production never eliminated custom-made systems, it largely overshadowed that older tradition. The shift in production methods profoundly altered the industry. Not only did the technology dramatically change, but also each kind of production involved a different set of social relations among technical experts, the company, and the customer.

The most notable change concerned the position of the engineer-designer. In custom production, the engineer's expertise was necessary for the design of every system that was made and sold. The company relied upon the intuitive leaps of experienced engineers to compensate for the lack of quantitative data in the new field. Because design expertise was a necessary part of competitive success, engineers had a strong position within the engineering firms that made up the majority of the early companies. The air-conditioning companies restrained the most costly flights of designers' fancy, but the self-interest of designers and their employers was closely allied. The company became a nearly transparent filter between the technical expert and the customer. Such was the power of the designer in custom-designed systems that one member of the industry called these installations "engineering ecstasy jobs."

The resulting product represented the specific needs of the customer, yet it also strongly reflected the rationality of engineering design and culture,

183

which emphasized performance, efficiency, and quantitative standards. Thus, custom-designed systems incorporated the values of the designers as much as the needs of the customer—an often overlooked aspect of such production methods. The close collaboration of engineers and customers so characteristic of custom design is often portrayed as technical expertise at the beck and call of monied interests, but designers were just as likely to view the buyer as a patron to finance an ideal engineering system. Indeed, in some cases, systems seemed to represent the designers' notions of proper technical solutions more than they did the customers' wishes.

Engineers knew that they largely controlled the air-conditioned environment under a system of custom design. Occasionally they exercised that power quite baldly, as did the designer of a school ventilation system who made it impossible for independent-minded teachers to provide heat without running the ventilation fan. And, just as clearly, the engineers were reluctant to share that power with others. The Chicago Ventilation Commission recommended that ventilation engineers refuse to equip cellar workshops with mechanical systems, so as to forestall the abuse of workers that might follow from management's monopoly over atmospheric conditions. The centralized control of environmental conditions that mechanical systems made possible seemed to many people to be a significant power indeed.

The strength of the technical experts with regard to the customer often produced an air-conditioning installation that subordinated all concerns to the technical requirements of the system. But the inflexibility of such a design led to conflict between air-conditioning firms and industrial customers. In this real push and pull between client interests and engineering standards, an articulation of the designers' vision for the new technology was important to narrow the gap between the two parties.

Conflict between engineers and clients was common. Engineers were frequently impatient with seemingly "irrational" users, for although technical experts could build their priorities into the design, once the system was installed, they were unable to control how the machinery was operated. The apparent concentration of power behind the designer was neither unquestioned nor unchecked. Factory and theater managers often disregarded design assumptions, good practice, and established standards; assumptions about both goals and methods might be swept away once installations were turned over to users. This seeming chaos resulted from the fact that in many cases those who negotiated and paid for the system were not those who would actually operate the equipment. In truth, the "consumer" was a generalized construct that on closer inspection resolved into at least two groups, buyers and users.

Air-conditioning companies constantly faced difficulties arising from this disjuncture. In one case, Carrier Engineering Corporation engineers were aware that they would have to "sell" the local management of Atlas Powder's Wolf Lake plant on a system that had been purchased by the national office but which the local management had not wanted. In another, CEC lost a contract because the operating personnel were familiar with a rival's controls and the buyer chose to capitalize on that familiarity. But such consideration of users' concerns seems sporadic in the custom-design industry. The buyer was represented best in the designer-consumer nexus, and all hell could break loose when the users gained control of the machinery.

Although users were the group least consulted in the design process, they retained considerable power over the machinery once delivered. The user who ran air-conditioning systems on 100 percent outside air, or 20 degrees colder than outside temperatures, or with 80 percent humidity rather than 50 percent was the bane of the custom-design industry. The technical community perceived such users as at best misinformed, and struggled to come to terms with the reality of their presence. Engineers could do little but proselytize.

But the informal challenges of consumers were just one manifestation of social conflict over the controlled indoor environment. That the indoor environment could be a scene of contested power is clear from the loosely organized groups that sought to use the state's regulatory power to break the hold of experts on the indoor environment. If the irrational user was the guerrilla fighter of those disenfranchised from the design process, the reformers were the regulars. Their challenge was fundamental, for they questioned the necessity of mechanical systems, the quantitative standards of performance, and the fidelity of designers' claims to equal or surpass the natural climate. The reformers' challenge had economic consequences as well, since it threatened a lucrative market for mechanical ventilation.

Many in the technical community believed that such conflict was mere misunderstanding and that the rationality of their solutions would be persuasive if properly explained. With such an assumption, they tended to see diehard critics as individuals in the grip of primordial passions beyond the discipline of rationality. Unfortunately, women were often used to exemplify such critics, and were portrayed as emotional and stubbornly obstructionist. The female spinners, nurses, schoolteachers, social workers, and housewives were depicted as standing in the way of a rational deployment of industrial systems, school plants, and residential installations. The decision by CEC to employ Margaret Ingels, a trained engineer, to promote comfort air conditioning among women's groups can be seen as one response to the perennial problem of educating away the conflict between engineers and such users.

The vision of man-made weather was both the rhetoric and the ideology of the air-conditioning engineer. It mediated the conflict between the engineering experts and the larger community—whether buyers, users, competing professionals, or regulatory agencies—by establishing a consensus through rhetorical persuasion. If the public agreed to the premise that artificial indoor climate was the goal, then the means to that end might more easily become an accepted necessity. But the analogy between air conditioning and natural climate encapsulated in the phrase *man-made weather* was more than a public-relations ploy—it was how engineers conceptualized their work. Engineers took natural climate as their model. Laboratory researchers, intrigued by the virtues of natural breezes on human skin, experimented with variable fans; impressed with the therapeutic effects of sunlight, they tested ultraviolet lamps. When the economics of comfort air conditioning dictated the use of recirculation, researchers took up the study of ionization as the key to nature's own constant refreshing of the atmosphere. Although they never acknowledged nature's superiority, engineers were nevertheless conscious of its appeal.

To the extent that the analogy between air conditioning and natural climate expressed the technical community's real aspiration, inspiration, and sense of accomplishment, it always contained two parts—the natural and the technical. Nature might be the model, but the standards were defined in the laboratory and expressed quantitatively. The use of such superlatives as "a June day at the beach" gave way to a concern with conditions within the comfort zone. Some more frankly imitative standards rather than ideal ones would have been possible, but ideal standards made nearly every natural climate appear inadequate and expanded the potential market for man-made weather.

The analogy between air conditioning and climate had far-reaching assumptions embedded in it, especially the notion that previously disparate technical functions were now swept up into one central mechanical system. Ventilation was wedded to temperature control and allied to humidity regulation. Catherine Beecher's ideal Victorian cottage, which deployed a window supply and a chimney exhaust for ventilation, fireplaces for heating, and a pot of water on the back of the stove for humidity, gave way to an air-conditioning system that incorporated all these functions. It was not the first time that mechanical systems were combined with traditional architecture. Sewers and plumbing, gas lighting, steam heating, elevators, and revolving doors all predated the introduction of air conditioning. Rather, air conditioning seemed a natural evolution in the transformation of the building into a machine. It was the modern building's iron lung.

Also inherent in this vision was the expectation that man-made weather now made it possible, desirable, and necessary to complete the separation of the outdoors from the indoors. The spray chamber brought humidity within engineers' control and provided the last essential element for reproducing natural climate. It was not only possible to produce artificial climate but highly desirable as well. The promoters of air conditioning argued that man-made weather was superior to the natural variety. Nature produced weather that was inconsistent, intractable, and compromised by society's industrial and urban development. Although air conditioning was modeled on natural climate, it was modeled on weather at its best, something even nature did not provide consistently. Seasonal variation was an expensive handicap for many industries, especially those with increasing year-round capital costs.

If the equation of air conditioning with man-made weather served to promote understanding and agreement between manufacturers and consumers, it also served to define the new technology among business rivals. It delineated the minimum functions of an air-conditioning system as control over temperature, humidity, cleanliness, and air circulation, for only such a comprehensive system could mimic outdoor climate. That definition of the proper functions of an air-conditioning system was in turn linked to specific designs.

CEC in particular pushed to define air conditioning in terms of both performance and a specific technology. For Willis Carrier, the two were intertwined. The control of humidity was at the heart of air conditioning, and he believed that the spray chamber yielded the most precise humidity control. For a time, standardization around spray chambers became one of the dominant themes of the company both within the engineering community and among consumers. Carrier seemingly saw little conflict between the standardization of air conditioning and the prosperity of his own company. His definition of air conditioning was one that privileged precise control in general, and his own firm in particular.

CEC also pursued what Mikael Hard has called the "scientification" of the marketplace, the use of science to tame unruly competition. Experimental science that established the primacy of humidity control for air conditioning ratified CEC's own technological choices and privileged the company over competitors whose air-conditioning systems did not provide for accurate humidity control. It was no accident that J. Irving Lyle provided the drive for the establishment of the ASHRAE laboratory and that Willis Carrier donated money to forward the research of the ventilation laboratory at the Harvard School of Public Health. While sensory standards of comfort let everyone be their own expert, the Comfort Chart set quantitative standards to guide both the industry and the consumer.

CEC's efforts to put the power of experimental science behind the analogy of air conditioning and artificial climate became more important than ever when the designers of process air conditioning expanded into comfort air conditioning. Theater managers and patrons were intensely interested in the new technology but willing to trade away the hallmark of industrial systems, precise humidity control. Although theater chains quickly bought into the appeal of artificial climate, precision was a luxury they were often unwilling to pay for. In the end it was not the public acceptance of precise humidity control that kept CEC at the top of the industry, but rather patent rights on the bypass system. This ensured that the company would make a successful transition to the comfort market. The bypass salvaged the spray chamber central to CEC's practice and at the same time resolved the tension between precision and affordability that was the largest stumbling block to expansion into comfort systems. Once that tension was resolved, CEC's design expertise netted it a comfortable share of these large-scale, custom-designed comfort systems.

Mass production changed the traditional configuration of this industry that had been characterized by precision, performance, engineering expertise, industrial oligopoly, conflict over consumer usage, humidity control, and the spray chamber. Indeed, there was little it did not change. Mass production brought with it instead a preoccupation with cost, salesmanship, patent pools, quantity production, consumer flexibility, refrigerated cooling, and fan-coil systems.

Nothing exemplifies changes in the air-conditioning industry better than the first appearance of the window air conditioner in the late 1920s. It was the antithesis of the old custom-designed systems. The new machines were sold largely on the basis of cost rather than performance. Although window units were manufactured in several sizes to accommodate different-size rooms, the match between cooling capacity and heat load was not calculated with the same precision as for central systems. It was hit or miss whether the units would bring conditions within the carefully established comfort zone, and they came with no atmospheric guarantees. But they did make environmental conditions slightly more comfortable, and many consumers settled for that standard of performance.

This dissociation of cooling equipment from the building was implicit in the transformation of air conditioning from a central system to a plug-in appliance. The lack of ductwork that tied the central system to its building was presented as a virtue to consumers during the 1930s. Window units were perfect for the renter, for the transient, for those stuck in traditional buildings that made no provision for bulky ducts. They were exceptionally flexible and adaptable to consumer needs. Indeed, the window unit became an adjunct to the consumer rather than to the building.

Residential air conditioning was built upon an established American business concept: mass production as the key to consumer affordability and business profits. This was especially important during the Depression, when some formula for business prosperity had to be reconciled with the consumer's diminished capacity to buy. Hopes for the widespread adoption of air conditioning rested on the ability of engineers to adapt the technology to small-scale applications, and on mass production to make the redesigned equipment affordable to large numbers of Americans. The shift in markets prompted a new technology, and the shift in markets and technology favored new companies.

Radio was the business model, refrigerators the technological model. Many companies had experience selling such consumer durables, and they fastened on air conditioners as a promising new product to carry them through the Depression. They expected to slot air conditioners into the established network of sales and distribution that already carried radios. Public utilities, electric manufacturers, and retail stores were all anxious to share in the prosperity that air conditioning promised.

The concern with cost, central to mass production, was felt not only in window units but also in residential systems. The widespread adoption of residential air conditioning took place in the postwar era, when a national concern for affordable housing for returning veterans sparked a speculative building boom. Tract development that combined new national mortgage policies, new construction methods, and standardized house design promised to solve the problem of affordable housing within the framework of commercial enterprise. Air conditioning benefited from all these cost-saving methods: Incorporating the cost of air conditioning into a thirty-year mortgage made systems easier to pay for; reducing the cost of ductwork lowered the initial price; and standardizing design lessened the expense of a custom-designed system.

Yet builders' concerns were largely confined to initial costs, and this meant that air conditioning underwrote a redesign of the American home. In speculative house construction, mechanical air-conditioning systems were often paid for by the elimination of passive cooling features that traditionally made the home environmentally comfortable. The inexpensive design of these tract homes depended on air conditioning to make them habitable and did less to hold down heat loads and operating costs. Air conditioning was often used as a means to an end—namely, making affordable designs habitable.

Using the mass-produced house as a vehicle for the spread of air conditioning offered considerable advantages. It made the technology affordable, and it reunited the machine and the building into a coherent package. The integration of technology and architecture generally provided better environmental performance for the customer. But once again, the buyer of the system was not the ultimate user of the installation. This led to a considerable gap between initial costs and operating costs. It was an important difference during the 1960s and 1970s, when the energy crisis in the United States provided a new guideline for judging technological design.

These three kinds of air conditioning—the custom-designed systems, the plug-in appliance, and the standardized installation for the tract home—illustrate the difficulties involved in integrating technical expertise into commercial products. None of these systems perfectly mixed the interests of the engineer, the company, and the aggregated groups known as the consumer. If the engineering culture that characterized the custom-design industry did not produce the best technology, neither did the market forces that dominated the mass-production industry. The custom-designed system produced the most technically rational design but, surprisingly, one that afforded the user little flexibility. The plug-in appliance privileged the consumer, but its complete divorce from the building compromised its performance. The standardized installation of the tract development provided affordability and performance, but in a building that was dependent upon its mechanical services and alarmingly inefficient in energy consumption.

How best to tap technical expertise remains an important issue. Altough engineers are just one of many groups that shape technology through its design and usage, they play an important role. If the vision of man-made weather seemed ultimately technocratic and confining, we still need to define a rational deployment of air conditioning—one that is technically sophisticated yet allows for a comprehensive and balanced set of values.

Notes

INTRODUCTION

1. Rosalyn Williams, *Notes on the Underground: An Essay on Technology, Society, and the Imagination* (Cambridge, Mass.: MIT Press, 1990).

2. Alex Witchel, "At Home with Horton Foote: Revisiting the Place He Never Left," *New York Times*, 6 April 1995.

3. Langdon Winner, "Do Artifacts Have Politics?" in *The Whale and the Reactor: A Search for Limits in an Age of High Technology* (Chicago: University of Chicago Press, 1986), 19–39.

CHAPTER 1

1. Bernard Nagengast, "John Gorrie: Pioneer of Cooling and Ice Making," American Society of Heating, Refrigerating and Air-conditioning Engineers (ASHRAE) *Journal* 33 (January 1991): S52, 54, 56, 58, 60–61.

2. See "Refrigerating by Street Mains from a Central Station," *Ice and Refrigeration* 1 (July 1891): 14–17; "Baltimore Refrigerating and Heating Co.'s Plant," *Ice and Refrigeration* 29 (July 1905): 1–18; David Branson, "Artificial Refrigeration through Street Pipe Lines from Central Stations," *Journal of the Franklin Institute* 87 (February 1894): 81–93.

3. "Living Room Refrigeration," *Ice and Refrigeration* 1 (September 1891): 148.

4. John J. Harris, "Cooling an Auditorium by the Use of Ice," *Heating and Ventilating Magazine* 4 (September 1907): 29–30.

5. R. Ogden Doremus, "Thoughts Suggested by Prof. Dewar's Discoveries," *North American Review* 156 (May 1893): 549–50.

6. Florence M. Wolff, "Alfred R. Wolff," *Stevens Institute Indicator* 26 (January 1909): 3–6.

7. A. P. Trautwein, "Professional Career of Alfred R. Wolff," *Stevens Institute Indicator* 26 (January 1909): 7.

8. Theodore Tenney to Donald Kepler, 22 April 1947, Box 28, Carrier Papers, Cornell University Archives, Ithaca, N.Y. (archive hereafter cited as Carrier Papers).

9. Richard A. Wolff to Margaret Ingels, 4 February 1949, Box 28, Carrier Papers. See also G. Richard Ohmes and Arthur K. Ohmes, "Early Comfort Cooling Plants," *Heating, Piping and Air Conditioning* 8 (June 1936): 310–12.

10. Alfred R. Wolff, "The Heating and Ventilation of Large Buildings," *Journal of the Franklin Institute* 138 (July–August 1894): 48.

11. "Developments in Heating and Ventilating during the Past Five Decades," *Heating and Ventilating* 15 (August 1918): 18.

12. "The Mechanical Plant of the Cornell Medical College, New York," *Engineering Record* 44 (October 1901): 324.

13. "Heating, Ventilating and Air Cooling at the New York Stock Exchange," *Engineering Record* 51 (April 1905): 436.

14. Ibid., 413.

15. The New York *Evening Journal,* as quoted in Ohmes and Ohmes, "Early Comfort Cooling," 310.

16. Wolff, "The Heating and Ventilation of Large Buildings," 122.

17. Ibid., 122.

18. "Meeting of the Engineers," *Ice and Refrigeration* 27 (July 1904): 17.

19. Wolff, "The Heating and Ventilation of Large Buildings," 46.

20. Trautwein, "Professional Career," 7.

21. "Heating, Ventilating and Air Cooling at the New York Stock Exchange," 413.

22. *Heating and Ventilating Magazine* 2 (September 1905): 26–27.

23. Quoted in Logan Lewis to Bob Fullerton, 17 December 1956, Box 4, Carrier Papers.

24. See Thomas R. Navin, *The Whitin Machine Works since 1831: A Textile Machinery Company in an Industrial Village,* Harvard Studies in Business History 15 (New York: Russell and Russell, 1969).

25. *New York Times,* 4 July 1940.

26. "Humidity in Textile Manufacturing," *Southern Textile Bulletin* (13 July 1911): 6.

27. *New York Times,* 4 July 1940.

28. U.S. Patent No. 811,383.

29. U.S. Patent No. 813,083.

30. U.S. Patent Nos. 821,989, 840,916, 840,917, 852,823.

31. U.S. Patent No. 852,823.

32. Interview with Willis Carrier by Margaret Ingels, 22 June 1948, Box 29, Carrier Papers.

33. *Weather Vein* 1 (June 1921): 9. The *Weather Vein* was published by CEC.

34. Margaret Ingels, *Willis Haviland Carrier, Father of Air Conditioning* (Garden City, N.Y.: Country Life Press, 1952). The published version of this biography was the result of Ingels's comprehensive draft being substantially rewritten by a company publicist. The original draft of her book can be found in Box 25, Carrier Papers (hereafter cited as Ingels, working manuscript, Carrier Papers).

35. Those firms included the Carrier Air Conditioning Company of America (1908–15 effectively), the Carrier Engineering Corporation (1915–30), and the Carrier Corporation (1930–   ).

36. "Henry W. Wendt," *Heating and Ventilating Magazine* 26 (July 1929): 102.

37. Ingels, *Willis Haviland Carrier,* 15–19.

38. Ibid., 26–27.

39. Willis H. Carrier, "Rational Psychrometric Formulae," American Society of Mechanical Engineers (ASME) *Transactions* 33 (1911): 1005. According to the *McGraw-Hill Dictionary of Scientific and Technical Terms,* edited by Sybil P. Parker, 5th ed. (New York: McGraw-Hill, 1994), *psychrometry* is "the science and techniques associated with measurements of the water vapor content of the air or other gases." A *psychrometer* is a device that makes use of two thermometers (dry- and wet-bulb) to determine "the moisture content or relative humidity of air or other gases."

40. Carrier, "Rational Psychrometric Formulae."

41. Willis H. Carrier and Frank L. Busey, "Air-conditioning Apparatus," ASME *Transactions* 33 (1911): 1055–1136.

42. Ingels, working manuscript, 192, Carrier Papers.

CHAPTER 2

1. *Weather Vein* 9 (March 1929): 72.

2. Robert S. Parks, "Fifty Years of Progress in Humidification and Air Conditioning," *Textile Industries* 112 (July 1948): 158.

3. H. W. Mason, "Air Conditioning Looks to the South," *Southern Plumbing, Heating and Air Conditioning* 1 (December 1945): 6.

4. American Moistening Co., *Automatic Humidity Controller* (Boston: American Moistening Co., 1912), 9.

5. Box 1, Carrier Papers.

6. Box 25, Carrier Papers.

7. Warren Webster and Co., "The 'Webster' Air Conditioning Apparatus" (Camden, N.J.: Warren Webster and Co., 1915), 11.

8. *Weather Vein* 2 (May 1921): 42.

9. Martin Campbell-Kelley, ed., *The Works of Charles Babbage, Vol. 8: The Economy of Machinery and Manufactures* (London, 1989), 85–86.

10. Frederick A. Halsey, *Methods of Machine Shop Work* (New York, 1914).

11. Joseph Wickham Roe and Charles W. Lytle, *Factory Equipment* (Scranton, Pa., 1935), 1. When they published their book, Roe and Lytle were both professors of industrial engineering at New York University. Roe's varied career had included stints at Worthington Pump and Machinery Co., Winchester Repeating Arms Co., Pierce Arrow Motor Car Co., and Yale University. He is best known today as the author of *English and American Tool Builders* (New Haven, Conn.: Yale University Press, 1916).

12. Roe and Lytle, *Factory Equipment,* 3.

13. Ibid.

14. Ibid., 9.

15. Ibid., 1.

16. L. L. Lewis, "Competition in the Early Days," 13 December 1956, Box 4, Carrier Papers.

17. Francis Thayer, "Solving Humidification Problems," *Textile World* (2 April 1927): 85.

18. Ingels, 27.

19. Ingels, working manuscript, 53–54, Carrier Papers.

20. Parks-Cramer Co., advertisement, *Southern Textile Bulletin* 17 (14 August 1919): 4.

21. Carrier's value was calculated from data published in W. H. Carrier, "A New Departure in Cooling and Humidifying Textile Mills," *Textile World Record* 33 (May supplement, 1907): 363–69. Cramer's value is cited in a memo by L. L. Lewis, "Competition in the Early Days," Carrier Papers. Lewis took Cramer's value from Stuart Cramer, *Useful Information for Cotton Manufacturers* (Charlotte, N.C., 1904). The presence of these data in print leads the historian to conclude that they were not secret. Nonetheless, abundant testimony from engineers on both sides—those privileged and those excluded—indicates that this information and a complex of values like it that were necessary for accurately determining heat loads were not generally known.

22. Lewis, "Competition in the Early Days," Carrier Papers.

23. Walter Fleisher, "Air Conditioning—Past, Present, and Future," speech presented at New York University, 1947, Box 28, Carrier Papers.

24. Stuart Cramer, U.S. Patent No. 311,383, filed 28 December 1904, issued 30 January 1906; Willis Carrier, U.S. Patent No. 854,270, filed 16 July 1906, issued 21 May 1907; William Braemer, U.S. Patent No. 885,173, filed 11 September 1907, issued 21 April 1908.

25. W. H. Carrier, "Progress in Air Conditioning in the Last Quarter Century," *Heating, Piping and Air Conditioning* 8 (August 1936): 447–59.

26. [A. P. Trautwein], "Record of the Principal Work of Alfred R. Wolff," *Stevens Institute Indicator* 26, no. 1 (January 1909): 14; "American Blower Corp.," Box 28, Carrier Papers.

27. "Topical Discussions: Advantages and Disadvantages Attending the Use of the Thumb Rules," American Society of Heating and Ventilating Engineers (ASH&VE) *Transactions* 12 (1906): 207.

28. Monte Calvert, *The Mechanical Engineer in America, 1830–1910: Professional Cultures in Conflict* (Baltimore, 1967), 6–7.

29. P. L. Davidson, "Air Conditioning Reminiscences," *Refrigerating Engineering* 37 (1939): 236.

30. Edwin Layton, "Mirror-Image Twins: The Communities of Science and Technology," *Technology and Culture* 12 (October 1971): 562–80.

31. "Proceedings, Twenty-third Annual Meeting, 1917: The Presidential Address," ASH&VE *Journal* 23 (April 1917): 348.

32. Ibid.

33. Ibid., 349.

34. J. D. Hoffman, "Reasons Why the Science of Heating and Ventilating Engineering Should Be Observed," ASH&VE *Transactions* 14 (1908): 89; "Topical Discussions No. 2: Reluctance to Divulge Alleged Secrets," ASH&VE *Transactions* 17 (1911): 309.

35. Buffalo Forge Co., "Buffalo Fan System of Heating, Ventilating and Humidifying," Catalog 700 (New York, n.d. [1920]).

36. Ingels, working manuscript, 188, Carrier Papers.

37. L. L. Lewis, "Memo Relating to Carrier Engineering Corporation and Its Agreement with Buffalo Forge Company," 13 November 1958; Lewis, "Fate Plays a Hand:

Another Month and We Would Never Have Organized Carrier Corporation," both in Carrier Papers.

38. "Carrier Research Service Available to Manufacturers," *Weather Vein* 5 (June 1925): 45.

39. L. L. Lewis, *The Romance of Air Conditioning* (Syracuse, N.Y., n.d.), 7, 9.

40. Lewis, "Competition in the Early Days," Carrier Papers.

41. "Architects and Engineers Manual Historical Notes," 20 June 1958, Box 28, Carrier Papers. Use of the phrases "we" and "our competitors" marks this unidentified memo as the product of a company member. Although it does not carry an attribution, it matches the style of many other memos that L. L. Lewis included in this collection and matches also the sentiments expressed in a similar document of known authorship.

42. A. M. Sanderson to J. I. Lyle, E. T. Murphy to A. E. Stacey, Jr., 6 December 1927, Stacey Correspondence, Box C-3, Carrier Corp., United Technologies Archives, East Hartford, Connecticut (hereafter cited as Stacey Correspondence).

43. L. L. Lewis to Henry Colton, 13 May 1955, Box 1, Carrier Papers.

44. "Carrier Service Available to Manufacturers," *Weather Vein* 5 (December 1925): 46–47.

45. E. T. Murphy to Carrier Engineering Co., Ltd., 24 February 1926, Box C-1, Stacey Correspondence.

46. John R. Allen, *Notes on Heating and Ventilation* (Chicago, 1905), 3.

47. "Architects and Engineers Manual Historical Notes," Carrier Papers.

48. L. L. Lewis, "Theatre Cooling," 7 September 1921, Confidential Data 24, Carrier Records, United Technologies Archives (archive hereafter cited as Carrier Records).

49. I. K. O'Brien to P. F. Pie, A. P. Shanklin to A. E. Stacey, Jr., 29 November 1933, Box C-1, Stacey Correspondence.

50. A. P. Shanklin to Chicago Office, 29 November 1933, A. E. Stacey Papers, Box C-1, Carrier Records.

51. A. P. Shanklin to A. E. Stacey, Jr., 4 December 1933, Box C-1, Stacey Correspondence.

52. H. B. Forbes to A. E. Stacey, Jr., n.d., Box C-1, Stacey Correspondence.

53. H. B. Forbes to A. E. Stacey, 27 August 1920, Box C-1, Stacey Correspondence.

54. P. L. Davidson to A. E. Stacey, 31 August 1934, Box C-1, Stacey Correspondence.

55. A. E. Stacey to C. G. Norton and E. T. Lyle, 5 March 1931, Box C-1, Stacey Correspondence.

56. Ibid.

57. E. P. Heckel, "Textile Humidities," 28 December 1920, Confidential Data 18, Carrier Records.

58. Ibid.

59. Parks-Cramer Co., advertisements, *Textile Bulletin* 48 (27 June 1935): 32; 48 (25 July 1935): 2; 49 (31 October 1935): 28; 50 (16 April 1936): 48; 50 (28 May 1936): 12; 50 (25 June 1936): 52; 50 (23 July 1936): 13; 50 (20 August 1936): 16; 51 (19 November 1936): 17; 51 (17 December 1936): 44; 51 (21 January 1937): 44.

60. *Textile Bulletin* 50 (28 May 1936): 12; 48 (27 June 1935): 32.

61. *Textile Bulletin* 48 (27 June 1935): 32.

62. Box 25, Carrier Papers.

63. "Philadelphia-made Macaroni Sent Abroad," *Confectioners and Bakers Gazette* 25 (February 1904): 26.

64. "Production of Macaroni," *Macaroni and Noodle Manufacturers Journal* 9 (August 1911): 7.

65. "Philadelphia Trade Conditions," *Macaroni and Noodle Manufacturers Journal* 5 (March 1907): 13.

66. "Greater New York and Brooklyn," *Macaroni and Noodle Manufacturers Journal* 5 (August 1907): 7.

67. Ingels, working manuscript, 213, Carrier Papers. In Ingels's manuscript the third company is identified only as "Foulds"; research strongly suggests it was Foulds Milling Co., Inc. See *Moody's Manual of Railroads and Corporation Securities,* vol. 3, Nineteenth Annual Number, Industrial Section (New York, 1918), 600–601.

68. R. T. Tree and P. L. Davidson, "Macaroni Drying with Ejector Unit Kiln," 30 October 1920, 6, Box C-12, Confidential Data, Carrier Records.

69. Ingels, working manuscript, 213, Carrier Papers.

70. Ibid., 214.

71. Ibid., 215.

72. Tree and Davidson, "Macaroni Drying with Ejector Unit Kiln," Carrier Records.

73. L. L. Lewis, 11 April 1949, Box 26, Carrier Papers.

74. L. Toro to Carrier Engineering Corp., 17 July 1920, Box C-1, Stacey Correspondence.

75. L. Toro to Willis Carrier, 17 June 1920, Box C-1, Stacey Correspondence.

76. This shift in factory production is comparable to Frederick W. Taylor's introduction of high-speed steel, which he also used to increase the influence of engineers in the factory, and not simply to strengthen the power of management.

77. "Discussion of Papers on Drying," ASH&VE *Transactions* 23 (1917): 279.

78. Ibid., 280.

CHAPTER 3

1. Albert W. Thompson, ed., *Air Conditioning in Textile Mills* (Fitchburg, Boston, and Charlotte: Parks-Cramer Co., 1924), 376–79.

2. G. B. Snow, "Refrigeration in Chocolate Mills," *Confectioners' and Bakers' Gazette* 31 (August 1910): 33.

3. Charles F. Hohmann, "Facts and Statistics," in William Augustus Evans, *Cellar Bakeries and Their Dangers to Producers and Consumers* (Chicago: Bakery and Confectionery Workers' International Union of America, 1910), n.p.

4. City of New York, Office of the Commissioner of Accounts, Raymond B. Fosdick, *A Report on the Sanitary Conditions of Bakeries in New York* (New York: M. B. Brown Printing and Binding Co., 1911), 9.

5. Robert P. Rasmussen, "Air Conditioning in the Baking Industry," *American Food Journal* (February 1928): 58.

6. American Blower Corp., *American Blower Air Conditioning Units,* Bulletin 6527 (1944), 1.

7. Margaret Ingels, *Willis Haviland Carrier, Father of Air Conditioning* (Garden City, N.Y.: Country Life Press, 1952), 38.

8. Willis H. Carrier, "Weather By Prescription," *National Safety News* (October 1938): n.p.

9. L. L. Lewis, "Dust—Notes on Prevention Thereof by Humidity and Circulation," 23 December 1919, Box C-12, Confidential Engineering Data 10, Carrier Records.

10. Ingels, *Willis Haviland Carrier,* 39.

11. Buffalo Forge Co., "Buffalo Fan System of Heating, Ventilating and Humidifying," Catalog 700 (New York, n.d. [1920]), 33.

12. Terry Mitchell, "Industry Finds Air Conditioning Indispensable," *Refrigerating Engineering* 25 (June 1933): 328.

13. Carrier Engineering Corp., advertisement, *Heating and Ventilating* 15 (June 1918): 85.

14. *Weather Vein* 1 (March 1921): 18–19.

15. *Manufactured Weather: The Equipment Developed and Utilized by Carrier Engineers,* Bulletin 58 (1929), 4.

16. Carrier, "Weather by Prescription."

17. Those contracts were with the American Ammunition Co. ($4,530) and the International Arms and Fuse Co. ($32,000). Ingels, working manuscript, Carrier Papers.

18. J. Irvine Lyle, "Effect of Air Conditioning upon Munitions Manufacture," ASH& VE *Transactions* 23 (1917): 384–85. It is unclear how many of these systems were installed by CEC.

19. "Fuse Loading," 19 June 1936, Carrier Documents, Box C-14, Carrier Records.

20. L. Logan Lewis, "A Philosophical Angle on Selling by J. I. Lyle Expressed as a Lecture to a Kid," June 1962, Box 7, Carrier Papers; Lyle, "Effect of Air Conditioning upon Munitions Manufacture," 383–92; J. Esten Bolling, "Refrigeration and Air Conditioning in the War," *Ice and Refrigeration* 57 (September 1919): 89–93.

21. For a general discussion of this technique, see Willis H. Carrier, "Recent Progress in Air Conditioning," *Refrigerating Engineering* 21 (March 1931): 187–89.

22. Memo, W. H. MacDonald, Jr., to L. L. Lewis, 4 November 1940, Box 30, Carrier Papers.

23. Ruttan Manufacturing Co., *Warming and Ventilation of Buildings* (Chicago: Press of America, 1888). Also, Ruttan Manufacturing Co., *Reports and Recommendations of Committees . . .* (Chicago: Shepard and Johnston, Printers, 1887).

24. E. V. Hill, "Ventilation Fundamentals and Test Methods," *Aerologist* 5 (June 1919): 50.

25. J. S. Billings et al., "Committee on Award in Sanitary Engineer School Competition" (8 March 1880). Reprinted in "Report of the Committee on School Room Ventilation," ASH&VE *Transactions* 18 (1912): 34. Billings had a noted career as an army sanitarian during the Civil War and later served as director of the New York Public Library. For his thoughts on ventilation, see J. S. Billings, *The Principles of Ventilation and Heating* (New York: Sanitary Engineer, 1884); and Billings, S. Weir Mitchell, and D. H. Bergey, *The Composition of Expired Air and Its Effects upon Animal Life* (Washington, D.C.: Smithsonian Institution, 1895).

26. William J. Baldwin, *The Ventilation of the School Room* (New York: By the author, 1901).

27. On the sewer-gas theory of disease, see Stanley K. Schultz and Clay McShane, "To Engineer the Metropolis: Sewers, Sanitation, and City Planning in Late Nineteenth-century America," *Journal of American History* 65 (September 1978): 389–411.

28. Passed April 1904 as an amendment to the Consolidated School Law with the active lobbying of state superintendent of public schools C. W. Cole, associate city superintendent of schools Andrew W. Edson, and members of ASH&VE.

29. New Jersey (1903), Pennsylvania (1905), Utah (1907), Virginia (1908), Vermont (1909), Indiana (1911), and North Dakota (1911) all enforced a standard of 30 cfm per person as the minimum rate of ventilation, as did the state boards of health in Massachusetts, Minnesota, and Connecticut.

30. Hill publicized his findings in Leonard Hill et al., *The Influence of the Atmosphere on Our Health and Comfort in Confined and Crowded Places* (Washington, D.C.: Smithsonian Institution, 1913); and Hill, *Sunshine and Open Air: The Science of Ventilation and Open-air Treatment* (London: H. M. Stationery Office, 1919–20).

31. For an account of his impromptu remarks, see "Discussion of the Papers of Dr. Burrage and Dr. Evans," ASH&VE *Transactions* 17 (1911): 141–46.

32. For the reception of Gulick's remarks, see "Report of the Committee on School Room Ventilation" (1912), 30. For an account of the formation of the committee, see "Committee on Schoolroom Ventilation," ASH&VE *Transactions* 17 (1911): 197.

33. D. D. Kimball, "Ventilation Problems," ASH&VE *Transactions* 18 (1912): 136.

34. Sherman C. Kingsley, ed., *Open-air Crusaders* ([Chicago]: United Charities of Chicago, 1910), 37.

35. As quoted, ibid., 68.

36. "Report of the Committee on Schoolroom Ventilation," ASH&VE *Transactions* 19 (1913): 107–8.

37. Hill et al., *The Influence of the Atmosphere on Our Health*, 2.

38. "Report of the Committee on Compulsory Legislation," ASH&VE *Transactions* 17 (1911): 17–18.

39. Kingsley, ed., *Open-air Crusaders*, 84.

40. See Sherman C. Kingsley and F. B. Dresslar, *Open-air Schools* (Washington, D.C.: Government Printing Office, 1917); Louise Dunham Goldberry, *The Open-air School and Out-door Education: A Resume, 1921* (Washington, D.C., 1921); "Potter Open Air School, Indianapolis, Ind.," American School Board *Journal* 70 (February 1925): 57–58.

41. Hill, *The Influence of the Atmosphere on Our Health*, 4.

42. New York State Commission on Ventilation (NYSCV), *Ventilation: Report of the New York State Commission on Ventilation* (New York: E. P. Dutton, 1923), vi.

43. "More about the New York State Commission's Report," *Aerologist* 1 (November 1925): 12–14.

44. George Whipple, "Students in School Ventilation, New York City" (1913), 1, Box 1, Whipple Papers, Harvard University Archives, Cambridge, Mass.

45. NYSCV, *Ventilation*, vii.

46. "Proceedings . . . 1917," 388.

47. Perry West to J. Irvine Lyle, 28 February 1918, J. I. Lyle file, ASHRAE Archives, Atlanta.

48. E. V. Hill, "How to Use the Anemometer," *Aerologist* 5 (April 1929): 19.

49. "Report of the Committee on Research," ASH&VE *Journal* 29 (March 1923): 109.

50. F. Paul Anderson and O. W. Armspach, "With a New Method of Making Air Dust Determinations," ASH&VE *Journal* 28 (July 1922): 542, 544.

51. *Weather Vein* 3 (1923): 7–8.

52. Ronald Grierson, *Some Modern Methods of Ventilation* (New York: D. Van Nostrand Co., 1917), 100.

53. F. Paul Anderson, "The Heating and Ventilating Engineer: A Tribute," *Aerologist* 3 (January 1927): 9.

54. Carrier Papers.

55. Jean Alonzo Curran, *Founders of the Harvard School of Public Health, with Biographical Notes, 1909–1946* (New York: Josiah Macy, Jr. Foundation, 1970), 161.

56. "Interesting People: Constantine P. Yaglou," *Aerologist* 3 (January 1927): 13.

57. "Interesting People: Dr. W. A. Evans," *Aerologist* 3 (June 1927): 13.

58. "The New York State Commission's Report: What Is Its True Value?" *Aerologist* 1 (September 1925): 2.

59. Charles-Edward Winslow, "The Dead Hand in School Ventilation," American School Board *Journal* 70 (June 1926): 143.

60. "Winslow's Window Complex," *Aerologist* 2 (March 1926): 12.

61. Editorial, *Aerologist* 2 (January 1926): 12.

62. For discussion of mechanical inventions modeled on human movement, see Siegfried Giedion, *Mechanization Takes Command* (New York: Oxford University Press, 1948).

63. Frank Hartman, "Ionization and Its Relation to Ventilation," *Aerologist* 5 (March 1929): 7.

64. Grierson, *Some Modern Methods of Ventilation*.

65. Edwin S. Hallett, "A New Method of Air Conditioning in School Buildings," *American City* 22 (April 1920): 423.

66. Grierson, *Some Modern Methods of Ventilation*, 3.

67. City of Chicago, Department of Health, Bureau of Sanitation, Division of Ventilation, *Report for 1911–1918* (Chicago, n.d.), 695.

CHAPTER 4

1. Willis Carrier, "How Air Conditioning Has Advanced the Art of Refrigeration," paper presented at the West Virginia section of ASME, 23 February 1943.

2. S. C. Bloom, "Trends in Air Conditioning," *Refrigerating Engineering* 30 (July 1935): 7. The best single article on early comfort air-conditioning systems is Bernard A. Nagengast, "The 1920s: The First Realization of Public Air Conditioning," ASHRAE *Journal* 35 (January 1993): S49–50, 52, 54, 56.

3. "The Editor's Page," *Heating, Piping and Air Conditioning* 10 (July 1938): 25.

4. Robert C. Allen, "Motion Picture Exhibition in Manhattan, 1906–1912: Beyond the Nickelodeon," in *The American Movie Industry: The Business of Motion Pictures*, edited by Gorham Kendem (Carbondale: Southern Illinois University Press, 1982), 14.

5. Ibid., 12.

6. Ibid., 14.

7. Commissioner of Accounts, "Report on the Condition of Moving Picture Show Places in New York" (March 1911), as quoted in "Report of Committee on Proposed Standards for Ventilation Legislation for Motion Picture Show Places," ASH&VE *Transactions* 19 (1913): 172.

8. City of Chicago, Department of Health, Bureau of Sanitation, Division of Ventilation, *Report for 1911–1918* (Chicago, n.d.), 745.

9. Ibid.

10. Ibid., 746.

11. As quoted in Allen, "Motion Picture Exhibition," 20.

12. "Popular Philadelphia House on Site of Nickelodeon," *Motion Picture News* 15 (9 June 1917): 3648.

13. *Chicago Daily Tribune,* 10 January 1914.

14. "Robinhood Theatre, Grand Haven, Mich., Distinctive for Its 'Simplicity,'" *Motion Picture News* 15 (10 February 1917): 955.

15. Charles Over Cornelius, "The Henry Miller Theatre," *Architectural Record* 44 (August 1918): 112–24, as cited in *Documents of American Theatre History: Famous American Playhouses, 1900–1971,* edited by William C. Young, vol. 2 (Chicago: American Library Association, 1973), 51.

16. Edwin A. Kingsley, "Heating and Ventilating a Theatre," in *American Theatres of Today,* edited by Randolph William Sexton, vol. 2 (New York: Architectural Book Publishing, 1930), 47.

17. "Whitehouse, Milwaukee, is 'Different,'" *Motion Picture News* 15 (7 April 1917): 2222.

18. "Typhoon Fans Go Well in South," *Motion Picture News* 15 (3 February 1917): 788.

19. Typhoon Fan Co., advertisement, *Motion Picture News* 15 (17 February 1917): 1121.

20. Ibid. See a correction to the price quote in *Motion Picture News* 15 (24 February 1917). Typhoon Fan Co., advertisement, *Motion Picture News* 27 (30 June 1923): 3207. The Typhoon Fan Co. equipped many theaters in 1917, including the Tivoli (Chattanooga, Tenn.), the Strand (Ridgefield Park, N.J.), the New (Fort Smith, Ark.), the Ampere (East Orange, N.J.), the Rivoli and the Harlem Strand (New York City), the Rialto (Waterbury, Conn.), the Columbia (Springfield), the Strand (Shelbyville, Ind.), the Old Mill (Dallas, Tex.), the U.S. (Bronx, N.Y.), and the Star and the Colonial (Savannah, Ga.). It also signed contracts with exhibitors in Jacksonville, Daytona, and Miami, Fla.; Norfolk and Newport News, Va.; New Orleans, La.; and Washington, D.C.

21. "Rivoli Makes Good," *Motion Picture News* (26 January 1918): 13.

22. "Directory of New Theatres," *Motion Picture News* 16 (11 August 1917): 1043.

23. O. W. Armspach, "Theatre Comfort: Temperature, Humidity, Air Motion . . . ," part 6 of 10, *Aerologist* 2 (May 1926): 17.

24. Typhoon Fan Co., advertisement, *Motion Picture News* 15 (17 February 1917): 1121.

25. Typhoon Fan Co., advertisement, *Motion Picture News* 16 (28 July 1917): 731.

26. Ibid.

27. "Old Mill Theatre, Dallas, Brought Up to the Minute," *Motion Picture News* 15 (23 June 1917): 3974.

200

28. "Successful Operation during Summer Months," *Motion Picture News* 28 (11 August 1923): 693–94.

29. "How Ventilation Helps Summer Business," *Motion Picture News* (16 June 1923): 2895.

30. Buffalo Forge Co., "Buffalo Fan System of Heating, Ventilating and Humidifying," Catalog 700 (New York, n.d. [1920]), 24.

31. Walter Fleisher, "Air Conditioning—Past, Present, and Future," 10, 12, Box 28, Carrier Papers.

32. Young, *Documents of American Theater History.*

33. A. E. Stacey to E. Bolling, 28 April 1920, Carrier Papers.

34. Margaret Ingels, *Willis Haviland Carrier, Father of Air Conditioning* (Garden City, N.Y.: Country Life Press, 1952), 147. United States Air Conditioning Corp. was founded in 1925 as a partnership between Walter Fleisher and the B. F. Sturtevant Co.

35. Buffalo Forge Co., "Buffalo Fan System," 24.

36. S. C. Bloom, "The Cooling and Ventilation of the Minneapolis Auditorium," *Refrigerating Engineering* 13 (April 1927): 298.

37. "Montgomery Gets One of the South's Finest Theatres," *Motion Picture News* 16 (20 October 1917): 2805.

38. Ibid.; Ingels, *Willis Haviland Carrier,* 142.

39. "Air Conditioning: Calculations for a Typical Theater," *Heating and Ventilating Magazine* 25 (February 1928): 114A.

40. David Naylor, *American Picture Palaces: The Architecture of Fantasy* (New York: Van Nostrand Reinhold, 1981), 47.

41. "'Mr. Barney' of the Balabans," *Chicago Herald-American,* 23 January 1941.

42. Lee Nusbaum, "Refrigeration in Air Conditioning," *Refrigerating Engineering* 7 (September 1920): 170.

43. "Fred Wittenmeier, Deceased," *Ice and Refrigeration* 64 (April 1928): 409; "Fred W. Wittenmeier," *Refrigerating Engineering* 15 (June 1928): 171. The first advertisement in *Ice and Refrigeration* for his new company appeared in the October 1918 issue (55:76).

44. Justin Christian Goosmann and Arnold H. Goelz were also instrumental in perfecting carbonic refrigeration in the United States. For Kroeschell Brothers Ice Machine Co., see J. C. Goosmann, "The Progressive Development of Carbon Dioxide Refrigerating Methods," *Refrigerating Engineering* 14 (December 1927): 188. For Goosmann's work, see "Charter Members of the American Society of Refrigerating Engineers," *Refrigerating Engineering* 28 (December 1934): 318; and J. C. Goosmann, "Factors Governing the Liquefaction of Carbon Dioxide," *Refrigerating Engineering* 14 (July 1927): 22. For Goelz's work, see Ingels, *Willis Haviland Carrier,* 132.

45. Fred Wittenmeier, "Cooling of Theatres and Public Buildings," *Refrigerating Engineering* 9 (October 1922): 115.

46. U.S. Patents No. 988,613 (26 October 1910) and No. 1,003,129 (18 May 1911).

47. U.S. Patent No. 988,613, 1.

48. Brunswick-Kroeschell Co., "Air Cooling and Conditioning for Public Buildings" (New Brunswick, N.J., 1920), 1, Box 2, Carrier Papers.

49. Wittenmeier, "Cooling of Theatres," 115.

50. Ibid., 116.

51. James S. McQuade, "Chicago News Letter," *Moving Picture World* (17 November 1917): 1021. Also "Central Park Theatre Opens in Chicago," *Motion Picture News* 16 (17 November 1917): 3512.

52. " 'Mr. Barney' of the Balabans."

53. Carrie Balaban, quoted in "Opening Night at the Tivoli," *Marquee* 11 (1979): 10–11. For further information on these theaters, see Naylor, *American Picture Palaces.*

54. Wittenmeier Machinery Co., advertisement, *Aerologist* 1 (December 1925): 7.

55. Bloom, "Cooling and Ventilation of the Minneapolis Auditorium," 298.

56. Brunswick-Kroeschell Co., "Air Cooling and Conditioning for Public Buildings." Recall that the cooling capacity of refrigerating machines is measured in the amount of heat absorbed by the melting of 1 ton of ice. To describe a refrigeration machine as a "250-ton machine," then, refers not to its weight but to its capacity to deliver in 24 hours the equivalent of the melting of 250 tons of ice.

57. "Orpheum Circuit: The Pioneer in the Evolution of Theatre Construction," n.d., n.p.; "State-Lake, Chicago," Theater Historical Society, Chicago. Those four imitators were the Hennepen-Orpheum (Minneapolis), the Mainstreet (Kansas City), the Golden Gate (San Francisco), and the Hill Street (Los Angeles).

58. Air conditioning was installed in the Los Angeles theater by Brunswick-Kroeschell; see Brunswick-Kroeschell Co., advertisement, *Aerologist* 1 (December 1925): 22. It was installed in the Kansas City theater by Wittenmeier Machinery Co.; see "Patron's Comfort First Considered in Mainstreet Theatre," *Motion Picture News* 25 (18 March 1922): 1657, 1660.

59. Everett S. Buck, "Theater Air Conditioning in the Southwest," *Heating and Ventilating Magazine* (February 1928): 74.

60. L. L. Lewis, "Air Conditioning in the Theatre," *Refrigerating Engineering* 14 (August 1927): 88.

61. Armand Carroll, "The Design of the Modern Theatre," in *American Theatres of Today,* edited by Sexton, 5.

62. Naylor, *American Picture Palaces,* 31.

63. Balaban and Katz, undated advertisement, "Chicago Theatre," Theater Historical Society, Chicago.

64. Randolph William Sexton, "Tendencies in the Design of the Present-day Theatre," in *American Theatres of Today,* edited by Sexton, 1–4.

65. Ben Schlanger, "The Theatre of Tomorrow," in *American Theatres of Today,* edited by Sexton, 52.

66. "Competition between Lobbies, Rather Than Pictures," *Motion Picture News* 15 (26 May 1917): 3339.

67. Naylor, *American Picture Palaces,* back jacket.

68. Leon Fleischmann, "The Theatre Owner and the Architect," in *American Theatres of Today,* edited by Sexton, 50.

69. Wittenmeier, "Cooling of Theatres," 117.

70. S. C. Bloom, "Air Conditioning," *Refrigerating Engineering* 13 (July 1926): 10.

71. Wittenmeier, "Cooling of Theatres," 118.

72. L. L. Lewis, "Air Conditioning's Contribution to the Picture Industry," *Boxoffice* (2 December 1944): 1.

73. Armspach, "Theatre Comfort," part 6 of 10, *Aerologist* 2 (May 1926): 17.

74. Lee Nusbaum, "Refrigeration in Air Conditioning," *Refrigerating Engineering* 7 (September 1920): 171.

75. Ibid., 116.

76. It is unclear whether this was the Tivoli Theatre or the Chicago Theatre.

77. Memo, New York office to Chicago office, 8 November 1919 (dictated 7 November 1919), Carrier Papers.

78. Ibid.

79. *"Auditorium v. Wallace:* Leo Logan Lewis Direct Examination," 1 (received 22 July 1940), Box 1, Carrier Papers.

80. U.S. Patent No. 1,583,060, filed 22 December 1924; reissued as U.S. Patent No. 16,611, filed 12 February 1927.

81. Naylor, *American Picture Palaces,* 85.

82. Charles Beardsley, *Hollywood's Master Showman: The Legendary Sid Grauman* (New York: Cornwall Books), 43, 67, 75. Grauman sold his share to Paramount Publix in 1925, and the theater was renamed the Paramount Theatre in 1929. It was demolished in 1960.

83. Quoted in Beardsley, *Hollywood's Master Showman,* 67.

84. *Weather Vein* 3 (December 1923): 36–37.

85. Bloom, "Trends in Air Conditioning," 7.

86. Quoted in Beardsley, *Hollywood's Master Showman,* 25.

87. For a description of the Million Dollar Theatre and the Egyptian Theatre, see J. H. Bailey, memo, 29 November 1956, Carrier Papers. The Million Dollar Theatre was reputed to be equipped with Buffalo Forge equipment, while the Egyptian Theatre had a Carrier air washer. For the Rialto Theatre, see "What a Complete Modern Ventilation System Does to the Air," *Motion Picture News* 26 (15 July 1922): 300.

88. Lewis, "Air Conditioning in the Theatre," 60.

89. [Willis H. Carrier], "Recent Progress in Air Conditioning," *Refrigerating Engineering* 21 (March 1931): 188.

90. Buffalo Forge Co., "Buffalo Fan System."

91. "San Francisco's Latest Creation, Granada Theatre, Opens Door to Public," *Motion Picture News* 25 (4 February 1922): 908.

92. Lewis, "Air Conditioning's Contribution," 1; Bloom, "Air Conditioning," 13.

93. Lewis, "Air Conditioning in the Theatre," 56.

94. Lewis, "Air Conditioning's Contribution," 1.

95. Ibid.

96. L. L. Lewis, "Theatre Cooling," Confidential Data 24 amended, Carrier Records.

97. Bloom, "Air Conditioning," 13.

98. Carrier, "Recent Progress," 188.

99. Walter Fleisher held bypass Patent No. 1,670,656, reissued as No. 20,088; L. Logan Lewis held patent reissue No. 16,611. Fleisher's patent rights appear to have been held by Cooling and Air Conditioning Corp., a partnership of Fleisher and established ventilation manufacturer B. F. Sturtevant.

100. Auditorium Conditioning Corp., *Why Only Air Conditioning Systems Licensed under the Patents of the Auditorium Conditioning Corporation Can Afford Dependable Perfor-*

mance, Scientific Effectiveness and Maximum Operating Economy in Auditoria, Theatres and Similar Structures (New York: Auditorium Conditioning Corp., 1929), 6.

101. Bloom, "Air Conditioning," 12.

102. Walter Fleisher, "Air Conditioning Old Buildings," *Refrigerating Engineering* 58 (December 1950): 1184.

103. Bloom, "Cooling and Ventilation of the Minneapolis Auditorium," 298–99.

104. Mikael Hard, *Machines Are Frozen Spirit: The Scientification of Refrigeration and Brewing in the Nineteenth Century—A Weberian Interpretation* (Boulder, Colo.: Westview Press, 1994).

105. Wittenmeier, "Cooling of Theatres," 118.

106. "The Aerologist," *Heating and Ventilating Magazine* 22 (November 1925): 81; Otto W. Armspach, "Air Conditioning Human Beings," *Refrigerating Engineering* 20 (December 1930): 351–53, 394, 399, 400.

107. Armspach, "Air Conditioning Human Beings," 351.

108. "Air Conditioning on Its Way," *Heating, Piping and Air Conditioning* 1 (December 1929): 696.

109. Malcolm Tomlinson, "Air Conditioning Movie Theaters," *Heating, Piping and Air Conditioning* 2 (November 1930): 935–36.

110. *Weather Vein* 9 (March 1929): 72.

111. Bloom, "Air Conditioning," 14.

112. Ibid., 13–14.

113. Maurice Olchoff, "Refrigeration in Air Conditioning for Comfort," *Refrigerating Engineering* 22 (September 1931): 164.

114. Bloom, "Air Conditioning," 14.

115. Ibid., 15.

116. "State-Lake," n.d., Theater Historical Society.

117. Willis H. Carrier and Leo Logan Lewis, "Status of Refrigeration as Applied to the Cooling of Theatres and Auditoriums in the United States" (1928), 7.

118. Memo, E. T. Lyle to A. E. Stacey and H. H. Pease, 9 September 1927, R-623, Carrier Records.

119. Tomlinson, "Air Conditioning Movie Theaters," 934.

120. P. L. Davidson, "Air Conditioning Reminiscences," *Refrigerating Engineering* 37 (1939): 242.

121. Leonard R. Phillips, "Air Distribution: All Important Phase of Air-Conditioning," *Architect and Engineer* 168 (January 1947): 11.

122. Tomlinson, "Air Conditioning Movie Theaters," 934–36.

123. Armspach calculated that equipment that provided 25 cfm with 75 percent recirculation was providing 6.25 cfm of outside air ("Air Conditioning Human Beings," 399).

124. Armspach, "Theatre Comfort," part 2 of 10, *Aerologist* 1 (November 1925): 7. Also Bloom, "Air Conditioning," 13.

125. Lewis, "Theatre Cooling Calculations."

126. City of Chicago, *Report,* 695–96.

127. Ibid., 699–702.

128. The department was aware of the existence of air-borne diseases and responded to the 1918 influenza epidemic by shutting down all theaters, lodges, and dance halls.

129. Samuel A. Ettelson, *The Chicago Municipal Code, 1922* (Chicago: T. H. Flood, 1922), 642–43.

130. "Municipal Control of Ventilation Practice," *Aerologist* 1 (October 1925): 3.

131. Bloom, "Trends in Air Conditioning," 31.

132. Tomlinson, "Air Conditioning Movie Theaters," 935.

CHAPTER 5

1. L. L. Lewis, "Defense Sales in World War II," sales conference, 12 December 1950, 4, Box 9, Carrier Papers.

2. L. L. Lewis, application for membership, ASH&VE, 29 August 1918, ASH&VE Archives, Atlanta.

3. "Weathermakers," *Fortune* 17 (April 1938): 122.

4. Willis B. Carrier, "Recent Progress in Air Conditioning," *Refrigerating Engineering* 21 (March 1931): 189.

5. W. J. Warren to L. L. Lewis, 23 March 1956; also L. L. Lewis, memo, 10 November 1955, and memo, 31 July 1956, Box 7, Carrier Papers.

6. New York City safety regulations adopted in 1915 required a licensed engineer for all installations exceeding 3 tons. For a history of the refrigeration safety code, see Harry D. Edwards, "The Refrigeration Safety Code: A Review of Two Decades," *Refrigerating Engineering* 28 (November 1934): 231–36, 240.

7. "What the Refrigerants Have Contributed," *Refrigerating Engineering* 28 (December 1934): 306.

8. *"Auditorium v. Wallace:* Leo Logan Lewis Direct Examination," 13, Carrier Papers. Lewis's testimony is in direct contradiction to Ingels's account, which describes the system as using carbon dioxide; Margaret Ingels, *Willis Haviland Carrier, Father of Air Conditioning* (Garden City, N.Y.: Country Life Press, 1952), 64. Lewis's account is reinforced by a memo written after an inspection trip, which describes "two 75HP horizontal Worthington feather valves, refrigerating compressors using ammonia as a refrigerant" located in the basement of the theater (J. H. Bailey, "The Grauman Theaters," 29 November 1956, 2, Carrier Papers).

9. Ingels, *Willis Haviland Carrier,* 74–75.

10. Ibid., 76–77. Also L. Logan Lewis, "Carrier's First Venture into Home Air Conditioning," 12 September 1962, 2, Box 7, Carrier Papers. The corporation was dissolved in 1936.

11. Lewis, "Carrier's First Venture," 2–3. The University of Illinois maintained a cooperative research center on warm-air heating from 1918 to 1970. For further information, see Seichi Konzo with Marylee MacDonald, *The Quiet Indoor Revolution* (Champaign, Ill.: Small Homes Council–Building Research Council, University of Illinois, College of Fine and Applied Arts, 1992).

12. Lewis, "Carrier's First Venture," 4.

13. L. L. Lewis to Stanley Klein, 18 November 1963, Box 8, Carrier Papers.

14. Carrier-Lyle Corp., 1929, Box 7, Carrier Papers.

15. Lewis, "Carrier's First Venture," 2, 5.

16. Ibid., 2.

17. Ibid., 3.

18. J. H. Ladew to Carrier-Lyle Corp., 20 April 1930, Box 7, Carrier Papers.

19. John Schmeig to Carrier-Lyle Corp., 3 January 1930, Box 7, Carrier Papers.

20. "Weathermakers," 118. Also "Carrier Corporation Purchases Plant," *Heating and Ventilating Magazine* (August 1928): 116.

21. "Carrier Corporation Purchases Plant," 116.

22. "Weathermakers," 120.

23. V. S. Day, memo, 7 October 1960, Box 7, Carrier Papers.

24. Lewis, "Carrier's First Venture," 10.

25. Ibid.

26. See Carroll Pursell, " 'A Savage Struck by Lightning': The Idea of a Research Moratorium, 1927–37," *Lex et Scientia* 10 (October–December 1974b): 146–61.

27. W. H. Carrier, "Air Conditioning in Relation to the Public Utilities," paper presented at the National Association of Power Engineers meeting, 29 August 1933, 1–2.

28. "Refrigeration and Material Progress," *Refrigerating Engineering* 25 (June 1933): 297.

29. Quoting Roger W. Babson in "The Editor's Page," *Heating, Piping and Air Conditioning* 10 (June 1938): 35.

30. "Industry Sees Air Conditioning as Eventual $5-Billion Market," *Business Week*, 16 March 1932, 11.

31. C. E. Stephens to L. Logan Lewis, 28 March 1930, Box 1, Carrier Papers.

32. As quoted in "The Popular Viewpoint on the Future of Air Conditioning," *Refrigerating Engineering* 28 (September 1934): 143.

33. "Domestic Air Conditioning Subject of Articles in Popular Magazines," *Refrigerating Engineering* 21 (1931): 276.

34. For a complete discussion of the technical difficulties in developing small residential air conditioners, see B. A. Nagengast, "Room Coolers prior to 1930 and the Technical Impediments to Their Development," ASHRAE *Transactions* 92, no. 2 (1986): 375–84.

35. David L. Fiske, "Air Conditioning and Engineering," *Refrigerating Engineering* 30 (July 1935): 25.

36. Everett R. Ryan, "How Will Air Conditioning Be Marketed?" *Refrigerating Engineering* 28 (August 1934): 65.

37. "Air Conditioning—In What Kind of Package?" *Refrigerating Engineering* 28 (August 1934): 61.

38. "Air Conditioning and Engineering," *Refrigerating Engineering* 30 (1935): 25.

39. "Domestic Air Conditioning Subject of Articles," 277.

40. "Air Conditioning—In What Kind of Package?" 61.

41. "The Editor's Page," *Refrigerating Engineering* 30 (1935): 313.

42. "The Editor's Column: Commercial Secrets," *Refrigerating Engineering* 28 (November 1934): 229.

43. U.S. Patent No. 551,107, filed 12 February 1895, issued 10 December 1895.

44. "New Rotary Ice Machine," *Ice and Refrigeration* 82 (February 1908); U.S. Patent No. 898,400, filed 20 June 1905, issued 8 September 1908.

45. "What the Refrigerating Machine Companies Have Contributed," *Refrigerating Engineering* 28 (December 1934): 304.

46. The Brunswick Refrigerating Co. also sold a remote unit for the home in 1913. See

Refrigeration in the Modern Home (New Brunswick, N.J.: Brunswick Refrigerating Co., 1913).

47. "A Review of Domestic Refrigeration," *Refrigerating Engineering* 25 (June 1933): 324.

48. "Audiffren-Singrun Refrigerating Machine for Ice Making and Refrigeration" (n.p.: H. W. Johns-Manville Co., 1913), 4.

49. J. F. Nickerson, "Development of Refrigeration in the United States," American Society of Refrigerating Engineers (ASRE) *Journal* 2 (November 1915): 79.

50. "Branch Association News," ASRE *Journal* 4 (July 1917): 71.

51. "Steps in the History of Refrigerating Systems," *Refrigerating Engineering* 21 (May 1936): 346.

52. "What the Refrigerating Machine Companies Have Contributed," 304.

53. Ibid., 358, 372.

54. "Home-made Weather Now Made at Home," *Business Week*, 5 November 1930, 13.

55. A 1950 history placed the introduction of the Frigidaire room cooler in 1928; see R. W. Morgan, "Room Air Conditioners—Past and Present," *Refrigerating Engineering* 58 (January 1950): 37. But others establish the date as 1929; see Nagengast, "Room Coolers prior to 1930," 375–84; and Nagengast, "The First Room Cooler," ASHRAE *Journal* 36 (January 1994): S60, 62, 64, 66, 68.

56. D. McCoy, "Frigidaire Air Conditioning," in *History of Frigidaire* (1964), quoted in Nagengast, "The First Room Cooler," S68.

57. Nagengast, "The First Room Cooler," S68.

58. "Home-made Weather Now Made At Home," 13–14.

59. General Electric Co. Refrigeration Division, "General Information on Air Conditioning," Box C-2, Stacey Correspondence.

60. A. R. Stevenson, Jr., to Brewster S. Beach, 15 June 1931; also General Electric Co., "General Information on Air Conditioning."

61. Mrs. John L. Kellogg to R. Cooper, Jr., 12 May 1931, Box C-2, Stacey Correspondence.

62. Stevenson to Beach, 15 June 1931.

63. Brewster S. Beach to A. R. Stevenson, Jr., 17 August 1931, Box C-2, Stacey Correspondence.

64. H. E. Snyder to A. E. Stacey, 1 July 1932, Box C-2, Stacey Correspondence.

65. L. L. Lewis, "The Evolution of Carrier's First Self-contained Room Cooler" (17 September 1959), I-6–8, Carrier Papers.

66. Paul Gant, U.S. Patent No. 1,860,357, filed 8 August 1928.

67. Lewis, "Evolution," I-1–6.

68. Ingels, *Willis Haviland Carrier*, 84.

69. "Weathermakers," 90.

70. Lewis, "Evolution," IV-1, II-4.

71. Ingels, *Willis Haviland Carrier*, 85.

72. Carrier Corp., Newark office, to Price and Data Book holders, memo, 12 September 1933, Box 7, Carrier Papers.

73. Esten Bolling, "Clients Will Demand Air Conditioning," *American Architect* 141 (May 1932): 45, 117.

74. Ingels, *Willis Haviland Carrier*, 85; sales bulletin, 27 July 1936, Carrier Papers.

75. F. H. Faust to Brewster S. Beach, 28 August 1931, Box C-2, Stacey Correspondence.

76. David L. Fiske, "The Year in Review," *Refrigerating Engineering* 30 (December 1935): 320.

77. "Air Conditioning Enters Mass Market, Launches New Industry," *Business Week*, 10 February 1932, 9.

78. "Competition Lines Up for Air-conditioning Sales," *Business Week*, 18 May 1932, 7–8. Also "Westinghouse Goes Hard after Air-conditioning," *Business Week*, 24 August 1932, 9.

79. "Air Conditioning," *Business Week*, 12 April 1933, 9.

80. "Home-made Weather," 14.

81. C. E. Greenwood, "The All-Domestic Team for 1932," National Electric Light Association *Bulletin* 19 (January 1932): 10.

82. C. E. Michel, "Review of the Utility's Position in the Air Conditioning Field," Edison Electric Institute *Bulletin* 4 (July 1936): 293.

83. Michel, "Review of the Utility's Position," 294.

84. See Nicholas B. Wainwright, *History of Philadelphia Electric Company, 1881–1961* (Philadelphia: Philadelphia Electric Co., 1961), 238; A. D. McLay, "Air Conditioning and the Central Stations," *Refrigerating Engineering* 25 (March 1933): 129.

85. "Air Conditioning: Third Selling Season," *Business Week*, 26 May 1934, 11.

86. Michel, "Review of the Utility's Position," 294, 314.

87. David L. Fiske, "The Year in Review," 319.

88. Michel, "Review of the Utility's Position," 293.

89. These four were awarded the John Scott Medal for invention. Additional names that appear on the patents are Alfred Weiland (U.S. Patent Nos. 2,112,870 and 2,148,596) and Arthur S. Locke (U.S. Patent No. 2,112,870).

90. H. K. Steinfeld, "Pioneer Developments in Self-contained Air Conditioning," ASHRAE *Transactions* 92, no. 2 (1986): 366–73; C. R. Neeson, "Room Cooler Design—Requirements for Self-contained Units and a Description of a New Machine," *Refrigerating Engineering* 26 (November 1933): 233–38; Henry C. Heller, U.S. Patent No. 2,113,691, filed 28 July 1934; Charles R. Neeson, U.S. Patent No. 2,131,355, filed 25 January 1933; also Henry L. Galson, U.S. Patent No. 2,052,561, filed 31 May 1933.

91. Charles R. Neeson, U.S. Patent No. 2,081,553, filed 13 February 1933.

92. Henry L. Galson, U.S. Patent No. 2,130,327, filed December 24, 1932.

93. Henry L. Galson, U.S. Patent No. 2,130,327, filed December 24, 1932; Charles R. Neeson, U.S. Patent No. 2,081,553, filed 13 February 1933; Hans K. Steinfeld, Henry L. Galson, and Arthur S. Locke, U.S. Patent No. 2,105,205, filed 25 February 1933; Alfred Weiland and Charles Neeson, U.S. Patent No. 2,112,870, filed 13 July 1933.

94. Steinfeld, "Pioneer Developments," 3.

95. "Air Conditioning—In What Kind of Package?" 61.

96. Quoted in Lewis, "Evolution," II-4.

97. Quoted, ibid.

98. John H. Holton to Thornton Lewis, 2 November 1931, Box C-3, Stacey Correspondence.

99. "Industry Sees Air Conditioning as Eventual $5-Billion Market," 11. Also "Air Conditioning," *Business Week,* 3 May 1933, 18.

100. "Air Conditioning, Twenty-five Years Old," *Business Week,* 5 December 1936, 15.

101. "Here Are Air-conditioning Facts," *Business Week,* 2 May 1936, 19–20.

102. "Air-conditioning Drive," *Business Week,* 1 June 1935, 11.

103. Willis Carrier, "What Makes and Keeps Our Jobs" (12 June 1940), "Refrigeration-Carrier," Division of Engineering and Industry, National Museum of American History, Smithsonian Institution, Washington, D.C.

104. "Weathermakers," 118.

105. Ibid. Aerofin was designed by Lawrence C. Soule in 1922. CEC purchased the Aerofin rights and shared them equally with B. F. Sturtevant, American Blower, and Buffalo Forge. However, the company retained exclusive manufacturing rights.

106. H. H. Mather, "The Economic Status of Air Conditioning," *Refrigerating Engineering* 31 (March 1936): 153.

107. H. A. Harer, "Plagues and Parasites of Air Conditioning," *Architect and Engineer* 130 (September 1937): 43.

108. "How the Refrigerating Industries Are Organized," *Refrigerating Engineering* 28 (December 1934): 337.

109. Michel, "Review of the Utility's Position," 293.

110. R. T. Brizzolara, "The Case for Ice," *Refrigerating Engineering* 30 (September 1935): 127.

111. S. C. Bloom, "Trends in Air Conditioning," *Refrigerating Engineering* 30 (July 1935): 8.

112. Brizzolara, "The Case for Ice," 144.

113. Union Ice Co., advertisement, *Architect and Engineer* 125 (April 1936): 5.

114. C. P. Yaglou, "Air Conditioning for Homes," *Journal of Home Economics* 26 (1931): 416.

115. Konzo and MacDonald, *Quiet Indoor Revolution,* 242–43.

116. The switch from steam-driven to electric-powered ice plants occurred in the early 1920s. As long as the ice industry used distilled water to produce a clear cake of ice, the exhaust steam from the steam engine could be used in the manufacture of distilled-water ice. The discovery of a process to produce ice from "raw water" helped usher in the change of power sources. See "Electricity Has Made the Small Ice Plant Possible," *Architect and Engineer* 74 (September 1923): 58.

117. Brizzolara, "The Case for Ice," 144.

118. Harer, "Plagues and Parasites of Air Conditioning," 43.

119. H. L. Lincoln, "Equipment for Summer Air Conditioning," *Architect and Engineer* 124 (June 1936): 51.

120. Frigidaire Co., advertisement, *Business Week,* 15 June 1932, 37.

121. "Various New Devices Offered as Home Air Conditioning Systems Are Developed," *Refrigerating Engineering* 22 (July 1931): 34.

122. "Air Conditioning, Third Selling Season," 15.

123. "Conditioning Code," *Business Week,* 5 October 1935, 12.

124. "It's Air Conditioning," *Business Week,* 30 November 1935, 23; "Here Are Air-Conditioning Facts," 20.

125. "It's Air Conditioning," 23.

126. Harer, "Plagues and Parasites of Air Conditioning," 43–44.

127. Ryan, "How Will Air Conditioning Be Marketed?" 94.

128. "Air Conditioning Starts Fast," *Business Week,* 1 June 1935, 11–12.

129. "Air-conditioning Drive," *Business Week,* 1 June 1935, 11–12. York Ice Machinery Co. moved into air conditioning around 1928 and became one of Carrier Corp.'s biggest rivals during the 1930s. Like Carrier, York emerged during the 1930s as both a manufacturer and a contractor. Some of the most lucrative large comfort installations of that decade were not movie theaters but the federal office buildings being constructed or renovated in Washington, D.C. Carrier Corp. air-conditioned the White House, the Capitol, the Supreme Court, the Department of Labor, and the Interstate Commerce Commission buildings. York Ice Machinery Co. installed equipment in the Post Office, the National Archives, the Library of Congress, the Department of Justice, the Senate Office Building Addition, and the Interior Department buildings. See Herbert Coward to A. E. Stacey, 18 March 1931 and 20 July 1932, Box C-5, Stacey Correspondence; Margaret Ingels, "Historic Air Conditioning," *Buildings and Building Management* (July 1939): 40–41; Ingels, working manuscript, 313–15, Carrier Papers; "Manufactured Weather Now Makes 'Every Day a Good Day' in the Nation's Capitol," *Weather Vein* 9, no. 3 (1929): 6–13, 35; "The National Capitol," *Heating, Piping and Air Conditioning* 2 (January 1930): 87; "What the Refrigerants Have Contributed," 307; "Air Conditioning Starts Fast," 11–12; Walter Fleisher, "Air Conditioning Old Buildings," *Refrigerating Engineering* 58 (December 1950): 1184–88, 1222.

130. Business Policy Letter 1500, 31 July 1931, Business Policy Letters—1931–33, Box C-6, Carrier Records. The Carrier factories were part of Carrier Manufacturing Corp., incorporated 9 April 1918 in New Jersey. On 15 April 1931 the Carrier Manufacturing Corp. was registered in Pennsylvania and the Allentown properties conveyed to that corporation.

131. "Weathermakers," 120. The five factories were the Carrier Plant at 750 Frelinghuysen Ave. and the Lyle Plant at 850 Frelinghuysen Ave., Newark, N.J.; the Brunswick Plant in Brunswick, N.J.; the Gamble Plant in Allentown, Pa.; and the Lewis Plant in Bridgeport, Pa. See Business Policy Letter 1500, 31 July 1931.

132. "Weathermakers," 120.

133. "Happy Trane," *Time,* 15 March 1937, 76.

134. "Here Are Air-conditioning Facts," 20.

135. Walter Fleisher, "Evaporative Cooling Comfort," *Refrigerating Engineering* 32 (December 1936): 413.

136. "Weathermakers," 87.

137. "Happy Trane," 77.

138. "Weathermakers," 87.

139. Ibid., 87, 120.

140. Ibid., 87.

141. "Air Conditioning" (3 May 1933), 18.

142. "Air Conditioning Twenty-five Years Old," 15. There were reportedly 2,494 air-conditioned railroad cars in the United States, of which 1,311 employed Freon, 224 used a steam-ejector system, and 942 were equipped with ice bunkers. See "What the Refrigerants Have Contributed," 307. These figures appear high for passenger cars, since one installa-

tion was made in 1929 and five more (ammonia systems) in 1930–31. The introduction of steam-ejector refrigeration around 1929 and the announcement of Freon in 1930 seem to have speeded up railroad air conditioning.

Railroad air conditioning made a contribution to the development of residential equipment both technically and by creating greater public awareness of the new technology. CEC engineer Hans Steinfeld was working on railroad air conditioning before he left the company to join the De La Vergne team working on self-contained room air conditioners. CEC set up an air-conditioned railroad car on a siding and invited 500 executives and financiers to inspect it, believing that few "men of money and influence" attended the movies, where the general public gained their acquaintance with air conditioning. See A. H. Candee, "Air Conditioning for Railway Passenger Cars," *Heating, Piping and Air Conditioning* 2 (December 1930): 1037–43; Rialto Cherne, "Developments in Refrigeration"; Steinfeld, "Pioneer Developments"; L. Logan Lewis, "A Thumb-nail Sketch of the Early Days of Air Conditioning and Carrier Composed Especially for the Complete Novice," 8 July 1960, Box 10, Carrier Papers.

143. Alice Thalman, "A Housewife Looks at Air Conditioning," *House and Garden* 70 (August 1936): 61.

CHAPTER 6

1. Refrigeration Equipment Manufacturers Association *Official Bulletin* 1 (May 1942): n.p.

2. Margaret Ingels, *Willis Haviland Carrier, Father of Air Conditioning* (Garden City, N.Y.: Country Life Press, 1952), 97; David L. Fiske, "Refrigeration in the War Years," *Refrigerating Engineering* 48 (December 1944): 468.

3. Ingels, *Willis Haviland Carrier*, 96; Ingels, working manuscript, 464, Carrier Papers.

4. See note 126 in chapter 5, above.

5. Loring F. Overman, "Washington News," *Heating and Ventilating Magazine* 41 (April 1944): 82; 41 (February 1944) 88.

6. Ibid., 41 (March 1944): 65.

7. Ibid., 41 (February 1944): 86.

8. Ibid., 41 (September 1944): 87.

9. John F. Haines, "The Economics of Air Conditioning: GI ACS," *Air Conditioning, Heating and Ventilating* 54 (May 1957): 89.

10. "ASRE Holds Air Conditioner Conference," *Heating and Ventilating* 51 (January 1954): 89.

11. T. W. Reynolds, "What's New That's Unusual," *Heating and Ventilating* (November 1952): 85.

12. Overman, "Washington News," 42 (October 1945): 71.

13. Loring F. Overman, "New Developments in Washington," *Heating and Ventilating* 42 (June 1945): 98.

14. R. W. Morgan, "Room Air Conditioners—Past and Present," *Refrigerating Engineering* 58 (January 1950): 38.

15. "Air Conditioning," *Life* 19 (16 July 1945): 39.

16. Willis H. Carrier, "New Prospects for an Established Industry," *Heating, Piping and Air Conditioning* 1 (May 1929): 30.

17. "Uses of Air Conditioners Grow," *Heating, Piping and Air Conditioning* 24 (January 1951): 121.

18. Ibid.; "Room Air Conditioner Sales Outlook Bright," *Heating, Piping and Air Conditioning* 28 (April 1956): 80.

19. "News of Equipment and Materials," *Heating and Ventilating* 49 (February 1952): 118; 49 (March 1952): 103, 106, 109, 110; 49 (April 1952): 120.

20. "Do We Make It or Buy It?" *Business Week*, 28 February 1953, 117.

21. "Air Coolers: Off on a New Spurt?" *Business Week*, 18 May 1957, 59.

22. Ibid.

23. David Hoppock, Burdge A. Gates, Richard H. Merrick, and Walter Grant, "Commercial Study of the Residential Air Conditioning Market," 30 September 1948, 24, Box 7, Carrier Papers.

24. William Bynum to Cloud Wampler, 6 May 1958, Cloud Wampler Correspondence, Box A, Carrier Records (hereafter cited as Wampler Correspondence).

25. E. P. Palmatier, "Residential Air Conditioning," *Heating and Ventilating* 51 (January 1954): 67–70; "Summer Cooling," *Air Conditioning, Heating and Ventilating* 52 (January 1955): 162–63.

26. Palmatier, "Residential Air Conditioning," 69.

27. "Residential Air Conditioning Discussed by ASRE," *Air Conditioning, Heating and Ventilating* 52 (February 1955): 114.

28. "Is This the Year for Built-in Cooling?" *Commercial Refrigeration and Air Conditioning* 12 (May 1955): 102.

29. "Air Coolers: Off on a New Spurt?" 59.

30. Bynum to Wampler, 6 May 1958.

31. R. A. Kroeschell to H. A. Stade, 8 June 1931, Box C-4, Stacey Correspondence.

32. Charles S. Leopold, "Department Store Cooling Installation," *Refrigerating Engineering* 31 (March 1936): 166.

33. Ingels, working manuscript, 299, Carrier Papers.

34. U.S. Patent File No. 16,611.

35. "Brief History of the Auditorium Conditioning Corporation," 1 November 1937, Carrier Papers; "Auditorium Patents," 1 November 1937, Box C-15, Carrier Records. These documents claim that a lawsuit was filed against Loew's and National Theatres Corp.; however, the patent file indicates only one lawsuit at this time, against Century-Parkway Corp. and United Kenomar Corp.

36. Carrier Corp. was by far the largest stockholder, controlling 40 percent.

37. *New York Times*, 29 December 1945.

38. "Room Conditioners Gain 50 Percent in Chicago," *Heating, Piping and Air Conditioning* 28 (September 1956): 65.

39. "Uses of Air Conditioners," 121.

40. "Air-conditioning Drive," 11–12.

41. "Air Conditioning for Homes," *Architectural Forum* 63 (August 1935): 136–37.

42. Overman, "Washington News," 43 (March 1946): 51.

43. Priorities Regulation 33, enacted in December 1945. Wyatt sought greater authority for government intervention from the legislature by supporting several of the provisions of

the Patman Bill (H.R. 4761), the Lanham Bill (H.R. 5455), and the Wagner-Ellender-Taft Permanent Housing Bill (S. 1592).

44. Overman, "Washington News," 43 (March 1946): 51.

45. Ibid.

46. Ibid., 43 (August 1946): 61.

47. "Small Home Plans Suggested for Veterans' Housing Program," *American Builder* 68 (August 1946): 78–79.

48. "Trends," *American Builder* 69 (August 1947): 19.

49. Ibid., 69 (September 1947): 18.

50. Frank W. Cortwright, "Cortwright's Column," *American Builder* 69 (September 1947): 37.

51. "The Case Study House Program," *Arts and Architecture* 62 (January 1945): 38.

52. Esther McCoy, *Case Study Houses, 1945–1962* (Los Angeles: Hennessey and Ingalls, 1977).

53. Case Study House One, despite its modest character, incorporated a spot room cooler. "Additional General Specifications for Case Study Houses One and Two," *Arts and Architecture* 62 (May 1945): 39.

54. *Architectural Review* (May 1959), as quoted in McCoy, *Case Study Houses.*

55. "The Case Study House Program," 4.

56. "100 Houses a Week," *American Builder* 68 (November 1946): 95–97.

57. Overman, "Washington News," 43 (October 1946): 63.

58. "Levitt and Sons Complete Veterans' Project," *American Builder* 68 (December 1946): 72–77, 154.

59. "The Building Outlook—1946 a Crucial Year—A Review and a Forecast," *American Builder* 68 (January 1946): 64.

60. Murray built 2,500 homes in Los Angeles and Orange Counties between 1940 and 1958. "Murray-Sanders Build Reasonable Priced Homes with Luxury Features," *Southwest Builder and Contractor* 131 (February 1958): 58–60.

61. See Archie Quincy Jones and Frederick E. Emmons, *Builders' Homes for Better Living* (New York: Reinhold Publishing, 1957).

62. "Complete Home Air Conditioning," *Fortune* 33 (April 1946): 260.

63. Frank Lopez, "Your New Home Can Be Designed for Air Conditioning," *Better Homes and Gardens* 27 (February 1949): 37.

64. "Plans and Home Plan Book," *Better Homes and Gardens* 27 (February 1949): 145.

65. Ibid., 40.

66. Lopez, "Your New Home," 38.

67. Westinghouse Electric Corp.'s Air Conditioning Division was wary of the changes, especially sealed windows. See Walter R. Yeary, "Problems in Residential Air Conditioning," *Heating and Ventilating* 50 (August 1953): 77–81.

68. "Complete Home Air Conditioning," *Fortune* 33 (April 1946): 260.

69. Ibid.

70. Servel Proof Book, Box 1, Collection 60, National Museum of American History Archives, Smithsonian Institution, Washington, D.C.; "Residents Sing Praises of Air-Conditioning," ARI *Koldfax* 7 (September 1956): 4.

71. "Briefly Stated," *Heating and Ventilating Magazine* 49 (October 1952): 80.

72. "Product Applications; Partial Conditioning Cuts Home Cooling Costs," *Heating and Ventilating Magazine* 49 (September 1952): 138.

73. "Air Conditioning Package for Low-cost Housing," *Architectural Record* 111 (June 1952): 217.

74. "The Surface Has Just Been Scratched," ARI *Koldfax* 3 (15 August 1952): 3.

75. "Home Community Air Conditioned," *Heating and Ventilating* 49 (February 1952): 110.

76. "News of the Month: Air Conditioned Homes," *Heating and Ventilating* 49 (October 1952): 140, 142.

77. Gilbert Burck, "The Air-conditioning Boom," *Fortune* 47 (May 1953): 203.

78. "Complete Home Air Conditioning," 260.

79. "Pushing Houses," *Business Week*, 24 January 1953, 31.

80. W. R. Woolrich, of the University of Texas, had been appointed by the Building Research Advisory Board of the National Academy of Sciences to a position of responsibility for promoting research in California, Texas, and Florida. The research was eventually contracted to the National Warm Air Heating and Air Conditioning Association under the direction of Warren Scott.

81. National Association of Home Builders (NAHB), Construction Department and Research Institute, *Preliminary Report on Comfort Cooling Performance, Air Conditioned Village Project*, Special Report 3 (19 January 1955); "Residential Air Conditioning Discussed by ASRE."

82. Department of Architectural Engineering of the College of Engineering, Bureau of Engineering Research, University of Texas, Austin, "Proposal for a Research Village Project," October 1953, A14 and A15, NAHB, Washington, D.C.

83. See Henry Wright, "Perfect Sleeping Temperature Year-round," *House Beautiful* 96 (January 1954): 42.

84. "A New Approach to Environment," *Architectural Forum* 106 (January 1957): 120.

85. Burck, "The Air Conditioning Boom," 203.

86. "Put Summer Cooling in the Package," *American Builder* 75 (May 1953): 82.

87. Jones and Emmons, *Builders' Homes for Better Living*, 203.

88. "You'll Have to Think Air Conditioning to Sell in '58," *American Builder* 79 (August 1957): 101.

89. Franklin Lynip to Cloud Wampler, 3 November 1958, "Advertising Dept.," Wampler Correspondence.

90. Cloud Wampler to William Bynum, 17 November 1958, Wampler Correspondence.

91. Cloud Wampler to William Bynum, 25 November 1958, "Advertising," Wampler Correspondence.

92. University of Texas, Austin, "Proposal," A6.

93. John E. Haines, "The Economics of Air Conditioning," *Air Conditioning, Heating and Ventilating* 54 (May 1957): 89.

94. "Office Conditioning Needed for Efficiency," *Heating, Piping and Air Conditioning* 22 (March 1950): 69.

95. *New York Times*, 26 June 1952, 1, 27; 27 June 1952, 1, 16; 28 June 1952, 16.

96. Cloud Wampler, "Growth of Air Conditioning Industry," *Air Conditioning, Heating and Ventilating* 54 (May 1957): 86.

97. "The Editors' Pages," *Heating, Piping and Air Conditioning* 29 (June 1957): 85.

98. "He Wouldn't Take 'No' for an Answer," *Commercial Refrigeration and Air Conditioning* 12 (April 1955): 107, 152.

99. "Survey Air Conditioning Needs of Government Buildings," ARI *Koldfax* 6 (September 1955): 8.

100. "News of the Month: Houston," *Heating and Ventilating* 49 (May 1952): 122.

101. "Office Conditioning Needed for Efficiency," *Heating, Piping and Air Conditioning* 22 (March 1950): 69.

102. Wampler, "Growth of Air Conditioning Industry," 86.

103. Nicholas B. Wainwright, *History of the Philadelphia Electric Company* (Philadelphia: Philadelphia Electric Co., 1961), 305.

104. Wampler, "Growth of Air Conditioning Industry," 86.

105. "The Editors' Pages," 29 (December 1957): 66.

106. "Mechanical Cost Second in Skyscraper Construction," *Heating, Piping and Air Conditioning* 28 (April 1956): 80.

107. "New Thinking on Office Buildings," *Architectural Forum* 99 (September 1953): 107.

108. "Air Conditioning Soaring Survey Shows," *Heating, Piping and Air Conditioning* 29 (April 1957): 126–27.

109. Carrier, "New Prospects," 29–30.

110. Esten Bolling, "Clients Will Demand Air Conditioning," *American Architect* 141 (May 1932): 44–45, 117.

111. "New Thinking on Curtain Walls and Window Sizes," *Architectural Forum* 99 (September 1953): 109–11.

112. "The Editors' Pages," 27 (August 1955): 82.

113. "A New Approach to Environment," 117.

114. "Block-type Designs, Lower Ceilings, Air Conditioning Urged for U.S. Buildings," *Architectural Forum* 104 (January 1956): 16.

115. "Government to Air Condition All New Buildings," ARI *Koldfax* 6 (December 1955): 6.

116. John E. Haines, "The Economics of Air Conditioning," *Air Conditioning, Heating and Ventilating* 54 (May 1957): 89.

117. "Government to Air Condition All New Buildings," ARI *Koldfax* 6 (December 1955): 1; "U.S. Government to Air Condition Most New Future Buildings," *Heating, Piping and Air Conditioning* 28 (June 1956): 91–92. "Effective temperature" is a combination of heat and humidity as measured by a dry-bulb and a wet-bulb thermometer.

118. "The Editors' Pages," 23 (July 1951): 67.

119. F. Morgan McConihe, "Policies and Plans of GSA in Air Conditioning Public Buildings," *Air Conditioning, Heating and Ventilating* 54 (May 1957): 97.

CHAPTER 7

1. U.S. Department of Commerce, Bureau of the Census, *1960 Census of Housing, Part I: United States Summary, Vol. 1: States and Small Areas* (Washington, D.C.: Government

Printing Office, August 1963), 39, table 7, I-28, table S, 39. There were 5,587,631 room units and 995,874 central systems.

2. Ibid., 39. Phoenix, Ariz., had 45,163 houses with central systems, compared to 23,841 households with one or more window air conditioners; for Las Vegas, Nev., the figures were 8,771 and 2,967, respectively; for Tucson, Ariz., 10,204 and 5,593; for Bakersfield, Calif., 6,912 and 5,805. All estimates based on a 15 percent sample.

3. Phoenix, Ariz. (24 percent), and Las Vegas, Nev. (22 percent).

4. Gallup and Robinson, Inc., "A Motivational Study of the U.S. Market for Central Residential Air Conditioning," vol. 1 (November 1959), 94, 97, 55, Carrier Records. Asked to rank the ten most appealing promotional features, 69 percent of respondents listed built-in kitchens, while 22 percent mentioned air conditioning.

5. Advertising Department, U.S. News and World Report, "Today's Customers for Home Air Conditioning" (n.p.: U.S. News and World Report, 1960), 18, 19, 25, 29. Similar occupations and income levels for owners were reported in another survey: 25 percent had incomes of $6,000–$9,999, and 50 percent had incomes over $10,000. Eleven percent refused to disclose their income. Gallup and Robinson, Inc., "A Motivational Study," 12, 13.

6. Gallup and Robinson, Inc., "Motivational Study," 12, 55. Thirty-three percent of nonowners cited expense as a deterrent; 7 percent mentioned operating costs.

7. Ibid., 27,, 55, 9.

8. Ibid., 52, 8–9.

9. George H. N. Luhrs Jr. Interview, 27, Arizona Oral History Collection, Hayden Library, Arizona State University, Phoenix.

10. *Phoenix Gazette*, 16 July 1934.

11. David W. Hoppock, Burdge A. Gates, Richard H. Merrick, and Walter Grant, "Commercial Study of the Residential Air Conditioning Market," 30 September 1948, 6–7, Box 7, Carrier Papers.

12. Walter L. Fleisher, "Air Conditioning Old Buildings," *Refrigerating Engineering* 58 (December 1950): 1184.

13. Central Arizona Light and Power Co., advertisement, *Phoenix Gazette*, 24 July 1934.

14. "Talk of the Town," *New Yorker*, 4 July 1959, 17.

15. National Association of Home Builders, Construction Department and Research Institute, *Preliminary Report: Comfort Cooling Performance Air Conditioned Village Project*, Special Report 3 (19 January 1955), 5.

16. ARI *Koldfax* 6 (April 1955): 1.

17. "Air Conditioning," *Life*, 16 July 1945, 39.

18. Schuyler Montgomery, "Round about Los Angeles," *Architectural Record* 24 (December 1908): 438.

19. Frank Williams, "Correcting California's Climate," *Sheet Metal Worker* 20 (October 1929): 641.

20. "Let's Talk 'Humiture,'" *Refrigeration and Air Conditioning Business* (August 1959): 52–53. The Discomfort Index was reputedly developed by Earl C. Thom, a climatologist with the U.S. Weather Bureau in Washington, D.C.

21. "Let's Talk," 52.

22. Gallup and Robinson, Inc., "A Motivational Study," 54.

23. Charles Leopold to G. Lorne Wiggs, 9 February 1951, Charles Leopold File, ASHRAE Archives, Atlanta.

24. NAHB, *Preliminary Report*, 3.

25. Sales conference paper, 18 September 1929, 6, Box 26, Carrier Papers.

26. Center for Research, *Consider the Air Conditioning Business as Your Career*, Careers Research Monographs 67 (Chicago, 1977), 12.

27. "Complete Home Air Conditioning," 260.

28. NAHB, *Preliminary Report*, 3.

29. Gallup and Robinson, Inc., "A Motivational Study," 55.

30. NAHB, *Preliminary Report*, 3.

31. "Russian Housing Experts Express Awe at Austin's Air-conditioned Homes," ARI *Koldfax* 7 (June–July 1956): 8.

32. Gallup and Robinson, Inc., "A Motivational Study," 32.

33. NAHB, *Preliminary Report*, 3.

34. Ibid., 5.

35. ARI *Koldfax* 6 (April 1955): 1.

36. Henry Wright, "Perfect Sleeping Temperature Year Round," *House Beautiful* 96 (January 1954): 42.

37. NAHB, *Preliminary Report*, 5.

38. That domestic technology in general raised household living standards rather than emancipating housewives is clearly argued in Ruth Schwartz Cowan's *More Work for Mother* (New York: Basic Books, 1983).

39. NAHB, *Preliminary Report*, 3–4.

40. Ellen Lupton, *Mechanical Brides: Women and Machines from Home to Office* (Princeton, N.J.: Princeton Architectural Press, 1993).

41. Jeremy Main, "A Peak Load of Trouble for the Utilities," *Fortune* (November 1969): 118.

42. "How They Keep Cool in the Desert States," *Business Week*, 24 January 1953, 160; " 'Swamp Coolers' May Find New Role on Central Valley Produce Farms," *Fresno Bee*, 30 December 1979.

43. Martin L. Thornburg and Paul M. Thornburg, *Cooling for the Arizona Home* (Tucson: University of Arizona, 1939).

44. Ibid., 15.

45. Michael Konig, "Phoenix in the 1950s: Urban Growth in the Sunbelt," *Arizona and the West* 24 (Spring 1982): 19–38.

46. Ibid. One California company was Mefco Manufacturing Co. of Glendale, which produced five sizes of evaporative coolers. "Evaporative Cooler," *Heating and Ventilating* 49 (October 1952): 121.

47. Konig, "Phoenix in the 1950s," 19–38. *Business Week* estimated one cooler for every three people in Phoenix. "How They Keep Cool in the Desert States," 160.

48. "The Thermometer Broke That Day," *Fresno Bee*, 22 May 1977.

49. "Grandpa's Ways to Beat the Heat," *Fresno Bee*, 22 May 1977.

50. "Air Conditioning Is a Star We Helped Create," *Fresno Bee*, 22 May 1977.

51. *Fresno Bee*, 22 May 1977.

52. "Air Conditioning Is a Star We Helped Create."

53. "Nobody Invented a Swamp Cooler, It Just Happened," *Fresno Bee,* 22 May 1977.

54. "Swamp Cooler Does Its Job at an Old-fashioned Price," *Fresno Bee,* 22 May 1977.

55. "Officials Cite Water Waste by Coolers," *Fresno Bee,* 29 July 1953.

56. One manufacturer of evaporative coolers estimated that in 1953 a 4,500-cfm evaporative cooler, big enough to cool a two-bedroom house, used $5 of electricity a month and $1 in water; in contrast, an air-conditioning unit used $30–$40 in power and $10–$15 in water. "How They Keep Cool in the Desert States," 160.

57. *1960 Census of Housing,* I-28; U.S. Department of Commerce, Bureau of the Census, *1970 Census of Housing* (Washington, D.C.: Government Printing Office), I-248.

58. "The Editors' Pages," 30 (April 1958): 81.

59. "Why Utilities Can't Meet Demand," *Business Week,* 29 November 1969, 49, 52.

60. *New York Times,* 13 July 1966.

61. Ibid., 14 July 1966.

62. Ibid.; Murray Schumach, "City Due for Relief Today after Nine Days over 90 Degrees," *New York Times,* 15 July 1966.

63. Arthur Daley, "Sports of the Times: Meet Me in St. Louis," *New York Times,* 12 July 1966.

64. *New York Times,* 13 July 1966.

65. Ibid.

66. "St. Louis a Leading City in Using Air Conditioning," *Heating, Piping and Air Conditioning* 24 (January 1952): 144–47.

67. *New York Times,* 14 July 1966.

68. Schumach, "City Due for Relief Today," 17.

69. *1960 Census of Housing,* table 26, 1–221.

70. *New York Times,* 16 July 1966.

71. "Calls Heat Up Repair Services," *Fresno Bee,* 4 August 1975.

72. "Air Conditioning Repairmen Busy," *Fresno Bee,* 14 June 1979.

73. "Insulation—An Idea Whose Time Has Arrived," *Fresno Bee,* 22 May 1977. A company representative estimated that most homes had R-5.5–R-7 insulation in the ceiling.

74. "Air Conditioning Is a Star We Helped Create." PG&E estimated the average cost of operating a 3-ton central air-conditioning system for the five-month season at $96 in 1977.

75. For example, during the drafting of a new building code in 1959, the director of planning and inspection first consulted with the Home Builders Association, Building Trades Council, Builders Exchange, PG&E, local chapters of the AIA, and professional engineers. Not all objections were incorporated in the final proposal. John Behrens to R. N. Klein, memo, 24 November 1959, "Uniform Building Codes, 1958–65," Legal: Codes, Fresno Municipal Records.

76. For an expression of these sentiments from the Fresno Citizen Participation Commission, see Eddie Mae Lomack to Fresno City Council, 7 January 1981, Fresno Municipal Records.

Essay on Sources

Two of the richest archival sources for a history of air conditioning are the Willis Carrier papers at Cornell University, Ithaca, New York, and the Carrier Corporation records at United Technologies, East Hartford, Connecticut. The Carrier papers include the working manuscript of Margaret Ingels's biography of Carrier, which is much more extensive than the published book, as well as charming and informative memos for the "Historian" from L. L. Lewis. The Carrier Corporation records, especially the A. E. Stacey correspondence, are a rich source for how engineers do their work.

Also helpful are the records housed at the National Museum of American History, Smithsonian Institution, Washington, D.C. The division records, catalog collection, and advertising collection yielded a variety of documents on specific companies.

The archives at the headquarters of the American Society of Heating, Refrigerating and Air-conditioning Engineers (ASHRAE) in Atlanta contains the records of both of its parent societies, the American Society of Refrigerating Engineers and the American Society of Heating and Ventilating Engineers. The archives preserve the presidential correspondence of many talented engineers and membership applications from a wider segment of the field.

Some papers of individual air-conditioning engineers have survived. The papers of Stuart Cramer are a story in themselves. Letters to Cramer from the company headquarters are preserved in the Whitin Machine Works collection at Harvard's Baker Library. Unfortunately, a set of almost daily correspondence between Cramer and Whitin executive George Marston Whitin from 1897 to 1910, which Thomas Navin drew upon for his study *The Whitin Machine Works since 1831: A Textile Machinery Company in an Industrial Village* (Harvard Studies in Business History 15 [New York: Russell and Russell, 1969]), is not in the inventory of the Baker collection. Whether that correspondence stayed with the company or is simply lost in the vastness of the Baker collection is unclear. All attempts to locate it at the library, the company, and the public library in Whitinsville, Massachusetts, failed. Cramer's wartime correspondence from Washington, D.C., is in the Southern Historical Collection, University of North Carolina, Chapel Hill. Alfred Wolff's correspondence with the building committee of the New York Stock Exchange is preserved in the library of the Exchange.

Turning to published sources, perhaps the most helpful book on the history of air conditioning is Margaret Ingels's "professional biography," *Willis Haviland Carrier, Father of Air Conditioning* (Garden City, N.Y.: Country Life Press, 1952). The chronological table at the back of the book contains some errors but is extremely useful because it gives a broader view of the industry than does the text. This reflects the character of Ingels's first-draft manuscript, which was heavily revised by a company publicist before publication; as noted above, the original draft is preserved in the Carrier papers at Cornell.

Some of the most careful research and balanced history writing has been done by Bernard Nagengast. Much of his work is gathered in Barry Donaldson and Bernard Nagengast, *Heat and Cold: Mastering the Great Indoors: A Selective History of Heating, Ventilation, Refrigeration and Air Conditioning from the Ancients to the 1930s* (Atlanta: ASHRAE, 1994). For those with less time, the well-written and nicely researched article by Robert Friedman, "The Air-conditioned Century" (*American Heritage* 35 [August–September, 1984]: 20–33), provides the best introduction. Raymond Arsenault's article "The End of the Long Hot Summer: The Air Conditioner and Southern Comfort" (*Journal of Southern History* 50 [November 1984]: 597–628) is one of the few pieces to focus upon the social impact of air conditioning.

The relationship of air conditioning to architecture is analyzed in Reyner Banham's insightful book, *The Architecture of the Well-tempered Environment* (Chicago: University of Chicago Press, 1969), which explores both mechanical and passive cooling. Several other articles also explore the house and its mechanical services: Eugene Ferguson, "An Historical Sketch of Central Heating, 1800–1860," in *Building Early America*, edited by Charles Peterson (Philadelphia 1976), 165–85; and Robert Bruegmann, "Central Heating and Forced Ventilation: Origins and Effects on Architectural Design," *Society of Architectural Historians Journal* 257 (1978): 143–60. On the special relationship between theaters and air conditioning, see Douglas Gomery, *Shared Pleasures: A History of Movie Presentation in the United States* (Madison: University of Wisconsin Press, 1992).

Nagengast has always been tactful in pointing out that although my own work critiques the efforts of heating and ventilating engineers to define the new technology, it necessarily places that group at the center of discussion and slights the contribution of refrigeration pioneers. For more material on the history of refrigeration, see the important series of articles by J. C. Goosman that appeared in *Ice and Refrigeration* between 1924 and 1927. Oscar E. Anderson drew upon Goosman's work for his seminal history of American refrigeration, *Refrigeration in America: A History of a New Technology and Its Impact* (Princeton, N.J.: Princeton University Press for the University of Cincinnati, 1953).

Mikael Hard's *Machines Are Frozen Spirit: The Scientification of Refrigeration and Brewing in the Nineteenth Century—A Weberian Interpretation* (Boulder, Colo.: Westview Press, 1994) is nominally about refrigeration, but it is more important for its portrayal of how business and technical elites build institutions and tap ideology in the establishment of technological paradigms. Also see Langdon Winner's "Do Artifacts Have Politics?" in

The Whale and the Reactor: A Search for Limits in an Age of High Technology (Chicago: University of Chicago Press, 1986), 19–39. Winner shows that technical design decisions "have politics"—that is, they establish an inherent set of social relationships around the use of the new technology.

This study emphasizes the influential role of engineers in the design and definition of new technology. There are several excellent works on engineers' institutional life: Monte A. Calvert, *The Mechanical Engineer in America, 1830–1910: Professional Cultures in Conflict* (Baltimore: Johns Hopkins Press, 1967); Edwin T. Layton, Jr., *The Revolt of the Engineers: Social Responsibility and the American Engineering Profession* (Cleveland: Press of Case Western Reserve University, 1971); Bruce Sinclair, *A Centennial History of the American Society of Mechanical Engineers, 1880–1980* (Toronto: University of Toronto Press, 1980).

The importance of production systems in shaping the power of the technical designer has been shown by David Hounshell, *From the American System to Mass Production* (Baltimore: Johns Hopkins University Press, 1984), and Phillip Scranton, *Figured Tapestry: Production, Markets and Power in Philadelphia Textiles, 1885–1941* (New York, 1989).

Finally, my interest in air conditioning was inspired originally by Sigfried Giedion's influential history of everyday technology, *Mechanization Takes Command* (New York: Oxford University Press, 1948). Giedion's concern with the mechanization of natural processes seemed especially applicable to the development of process air conditioning.

Index

air conditioner: as appliance, 127–29; costs, 143; definition of, 124; development of, 129–30; popularity of, 147

air conditioner, window, 3, 5, 111, 140, 178, 188–89; in offices, 159; peak sales, 179

air conditioning: adoption of, 166; affordable systems, 112, 143, 144; buyers, 184–85; condensers, 122; control instruments, 20; costs, 124, 138, 144, 152, 162, 190; definition of, 135–36; ductwork, 189; markets, 110–11; ownership, 167; partial systems, 112; residential, 5–6, 113; retrofitting, 147; standardized design, 190; statistics, 138; users, 184–86. *See also* air distribution; air washer; cooling; process air conditioning; ventilation

air distribution: bypass, 101–2; downdraft, 100; grilles, 100; mushroom ventilators, 99–100; upward method, 99–100

air washer, 25–27, 55, 113; for evaporative cooling, 87

Allen, John R., 40, 69, 70

American Blower Corporation, 33, 54, 88; and bypass patents, 146; controls, 35; pioneering role, 36

American Society of Heating and Ventilating Engineers (ASH&VE), 60; and comfort air conditioning, 104; formation of, 10; presidential address (1917), 37; rejection of fresh air, 107–8. *See also* ASH&VE Research Laboratory

American Society of Mechanical Engineers (ASME), 27, 42

American Society of Refrigerating Engineers (ASRE), 69, 122

Anderson, F. Paul, 69–71, 74

Armspach, Otto, 86, 103

ASH&VE Research Laboratory, 68–69, 73, 101–2, 188, 115

Atlas Powder Company, 43, 185

Auditorium Conditioning Corporation: challenge to, 145–46; dissolution, 146; dominance, 102; incorporation, 101

Austin Air Conditioned Village: establishment of, 154–55; experts and, 173; hot-weather lifestyles, 170; housekeep-

Johns Hopkins Studies in the History of Technology (New Series)

LIBRARY OF CONGRESS CATALOGING-IN-PUBLICATION DATA

Cooper, Gail, 1954–
    Air-conditioning America : engineers and the controlled environment,
1900–1960 / Gail Cooper.
        p.   cm. — (Johns Hopkins studies in the history of technology ; 23)
    Includes bibliographical references and index.
    ISBN 0-8018-5716-3 (alk. paper)
        1. Air conditioning—United States—History.   I. Title.   II. Series:
Johns Hopkins studies in the history of technology ; new ser., no. 23.
    TH7687.5.C66   1998
    697.9′3′0973—dc21   97-37148
                                                                          CIP